Pigs

To Jackie The World's #1 Pig Lover

May This Book Bring You Many Hours of Fun 'n Laughter!

Love, Terry

Pigs

Sarah Bowman and Lucinda Vardey

A Nicholson Press Book

MACMILLAN PUBLISHING CO., INC.
New York

Macmillan Publishing Co., Inc. 866 Third Avenue, New York, N.Y. 10022

The Nicholson Press
9 Sultan Street
Toronto, Ontario M5S 1L6

First American Edition 1981

Designed by Maher & Murtagh
Edited by Barbara Purchase

Printed in the
United States of America

Library of Congress Cataloging in Publication Data

Bowman, Sarah
Pigs

"A Nicholson Press book"

Summary: Uses a variety of literary forms to celebrate the pig.
    1. Swine — Literary collections.
    2. Swine — Addresses, essays, lectures.
    [I. Pigs — Literary collections]

1. Vardey, Lucinda. II. Title.
PN6071.S94V37   810'.8'036 81-3757
AACR2
ISBN 0-02-514140-6

*For our parents
Elspeth and Jack Bowman,
Edwina and Lew Vardey,
with thanks for days in clover.*

Sarah and Lucinda wish to thank the following people for their help and support in compiling this book: Nancy Cunningham for moral and mechanical support; George Krzeczunowicz, the wonder warthog, for being a true piggy person; Jill Shefrin from the Osborne Collection for her time and inspiration; Andrew Thomas, whose unique picture collection keeps the pig alive; Edwina Vardey for ceaseless and patient research in unearthing forgotten treasures; Lewis Vardey, whose magic never fails; Suzanne Stormont for her pink meringues; Allan Stormont and Patrick Crean for flying with us, and Keith Abraham and Peter Maher for wallowing with us.

Also thanks to Penelope Anderson, Elspeth Bowman, Patti Bragg, Carolyn Brunton, Nancy Colbert, Louise Dennys, Liz Franklin, Timothy Geary, Rena Grant, Betty Henderson, Manya Igel, Bill Kime, Eric Koch, Stefan Krzeczunowicz, Greg Lawson, Alberto Manguel, Robert Russell, Jenny Scazighino, W. Skarbek-Borowski, Michael Stephen, Shizuye Takashima, Ann Vanderhoof, Anne Wheatley-Hubbard, Grahame Woods, and Sophia Yurkevich.

A special mention to Erik Peters, who ate, drank, and dreamt of pigs for well over a year, and to our editor, Barbara Purchase, whose companionship, patience, and grammatical knuckle-rapping finally got pigs on to the printed page.

Contents

Talking About Pigs. . . .

"**W**hat is it that sums up that all-round feeling of security you get when you see a contented pig nestling in its sty?"

"Affinity?"

"That's it. Affinity. Man and pig have shared an affinity for centuries."

"And people who breed pigs today literally glow when asked to pinpoint what's special about the animal. There's a definite respect tinged with amusement, and a grudging admiration for the family *suidae*."

"I can remember my mother always stopping the car and pointing excitedly out of the window whenever we passed a field of pigs."

"My introduction was more accidental, in fact. It was on a godforsaken station platform somewhere in northern Italy — piglets were all crammed into a crate, and I still remember how they squealed and grunted through the cracks. Then, it was a huge pink sow, Agatha, who became the pig of my childhood. She belonged to some farming nuns up the road, and tea time there was never complete without spilling the leftovers into her trough."

"The foundation of our love for pigs definitely took root when we were young, although that child-like affinity still remains now that we're older."

"Let's say it's more of an urban affinity now — a touch of the barnyard tinged with red brick."

"But why are there so many of us who feel the same way about pigs? Just think about all the incredible piggy people we know."

"Yes, pig fanatics turn up everywhere and always unexpectedly. When you tell them you're going to write a book on pigs, they react with 'marvelous' or 'wonderful,' and then proceed to tell you a hilarious story involving a pig and their Aunt Maud, or else they introduce you to another piggy person."

"Exactly. I would never have found Mr. Andrew Thomas's Art Gallery in Oxfordshire if it hadn't been for a piggy man in Toronto, and I would never have met Anne Wheatley-Hubbard, who owns the oldest registered herd of Tamworths, if I hadn't mentioned pigs at a tea party in Wiltshire. The next thing I knew, I was over at Anne's farm looking at a palace full of pigs, including a very lively character called Maple, who is actually descended from Tamworths in Canada. But the most unforgettable sight wasn't in the barn. In Anne's hallway, four massive cabinets reaching from floor to ceiling overflowed with at least 1200 'pigs' cast in different shapes and sizes; silver, bronze, china, and even straw pigs from every corner of the globe."

"It must be the largest pig collection in the world."

"It dwarfs mine drastically, and it proves a point: there's something special about the animal."

"It's partly the look, I think. That rosy, rotund body etched with bristles, the trim trotters, and that beady glint in the eye. Even the tail has individuality. And, of course, it's the personality."

"Independent, resourceful, unpredictable, and confident. There's also a wry sense of humor behind that rueful glance."

"You've got it. No wonder some writers have been captivated by the pig."

"And artists, too."

"Well, here's our book. People and pigs: the unique relationship."

"People and pigs. . . .It's a traditional theme that goes back to the cavemen and runs through the boar hunts

of history."

"It reaches great heights in mythological and religious symbolism, and yet plunges to great depths in ridicule and satire, and in witchcraft and taboo."

"But it's a theme that ends up in tranquil domesticity — a well-cared-for pig peacefully embedded in its sty. A sublime prospect!"

"It could be a sobering and sad one, too."

"Why?"

"Well, our grandchildren may never hear the grunt or witness the wallow. Mass production doesn't only apply to T-shirts. Factory farms are here to stay. Pigs are now being fattened up under artificial conditions before heading for the conveyor belt."

"We'd better hurry up then, so that a vestige of today remains — even if it's only a pig on a printed page."

S.B. & L.V.,
May, 1981, Toronto

THE TAPIR

THE HEDGEHOG

THE PIGLET

THE SHOAT

THE RIG

THE SOW

LEWIS VARDEY '81

The Gilt

THE HIPPOPOTAMUS

THE PORCUPINE

THE WARTHOG

THE PECCARY

THE BOAR

THE BARROW

Pigstory — A Brief Glimpse

It is hard to believe that the pig as we know it today has not changed much in thirty-six million years. While other, surly beasts roamed planet Earth (to their eventual extinction), pigs darted through the thickets and forests, happily living off the nuts and grain of the land.

The relationship between man and pig touched off roughly around the time that Neolithic man appeared on the scene. Pigs either found themselves fighting him or living with him, and those not caught by him in the flesh were captured in spirit on the cave walls.

The earliest record of pigs as domesticated animals dates back to 5000 B.C., when the Chinese introduced pigs into their homes — though what they both did within is entirely left to our imagination. Whatever it was, there can be no doubt that the pigs at least provided entertainment; their wit is undeniably

one of their best attributes, but, unfortunately, it has not always been readily appreciated. The French monarch Louis XI is known to have ordered pigs to dance for him while he lay around his palatial bedroom. One could speculate that his motives were purely spiteful (had he just been outwitted by a boar while hunting in his chateau grounds?),

especially as the dancing troupe was forced to adorn pantaloons and ribbons and to perform to the subtle accompaniment of bagpipes.

In England, however, during the eighteenth century, performing pigs were accorded more respect. At county fairs and street stalls, passers-by paid to stand in line and witness "learned pigs" picking out cards,

ringing bells in time to tunes, and generally clowning around. Yet as late as the mid-nineteenth century, there were still instances of pig mistreatment. This particularly painful report is taken from *Musical World*, November 1839:

The Porco-Forte is the name of a new musical instrument said to have been invented in Cincinnati, of course. It is a long box divided into compartments, one for each note, for as many octaves as may be wished. Into each division, a pig is placed, and the tails of the porkers run through holes in the side of the box, arranged like the keys of a piano. The tails are pinched by a sort of spring and lever machinery, and the effect is said to be delightful. If the pigs are well selected, they will wear about three years without tuning. (Ouch!)

Offshoots of the family *Suidae*

The Chinese

The Peccary

The Hampshire

The Berkshire

Pigs were put to more practical use as farming aids. Among their skills was the sowing of seeds, which they trod into the earth after the rain had softened the ground. So important was their function, in fact, that in ancient Egypt an office of Overseer of the Swine was created by King Sesostris I.

Unlike other farm animals, pigs are omnivorous and so are not easily herded to graze. They require a dwelling set apart: that secure enclosure — the unmistakable, irresistible sty. But sties were not always available, especially in medieval times, when pigs made it their business to run all over and root up the land of lordly landowners. A law was passed stating that instead of making meals out of scraps in the streets, pigs were to be fed in the homes of their owners and were to wear snout rings as a further deterrent to crop destruction. Peasants took to keeping two pigs — one to fatten and later consume, and one to have as a pet. (What a cosy atmosphere there must have been in those rural kitchens!)

The Old Irish Greyhound Pig (After an illustration by Richardson c. 1850)

All sorts of species of pigs emerged, notably the Chinese breed, a popular strain with European traders. The Italians introduced a pig appropriately dubbed the Neopolitan, while England nurtured many breeds, most named after prominent counties such as Yorkshire, Berkshire and Suffolk. The Welsh had their Welsh Pig, and the Irish had a particularly unique variety called the Old Irish Greyhound Pig.

It took some time before the pig found its true venue in the farming community. The French "chercheuse," the inimitable detective of the strong snout, proved to be invaluable in uncovering that exotic subterranean fungus, the truffle:

Because the truffle grows underground, humans need animal help to hunt it. Périgourdins use pigs, preferably the more docile sows, who have a keener sense of smell than male pigs. When a Périgourdin goes to buy a chercheuse (truffle-searching sow), he first determines a price with the farmer. Then, secretly, he drops a small truffle in the pigpen and crushes it under his heel. If the pig gets excited, he buys her — for the price quoted. For the farmer, it's too late to raise the price in the light of the sow's predilection for truffles. A sow is a chercheuse till she's five; then she loses her sense of smell and retires to await her fate as bacon.

Out in the woods, the sow sniffs and grunts with all the passion of a foxhound on a scent. She runs to and fro and then suddenly starts digging. The hunter rushes over and digs that spot with a small spoon-like utensil. Nine out of ten times, it's a truffle, and if he didn't hurry the sow would gobble it. He gives the faithful sow maize, chestnuts, or acorns to distract her. On very good hunting days, she might get a small truffle to console her.

Truffles ripen between October and March. They never grow in bunches, though they're always within three yards of their host trees. If the tree is hurt, the truffles rot — so hunters take great pains to keep their sows from doing nature's duty near the oaks. The Périgourdins are close-mouthed people. They often hunt truffles with flashlights at night to keep their truffle grounds secret. And it's not uncommon for them to prefer the privacy of their mattress ticking to the bank as repository for thousand franc bills. They treat their chercheuses with affection. If someone invented a truffle Geiger counter, a hunter would probably buy it, but he'd miss his pet.

from *Truffles:*
An Underground Delicacy,
Joanne Kates

Although we, the authors, beg to differ, the dog is commonly referred to as "man's best friend." Yet the sad fact that dogs are now being trained as truffle hunters cannot be overlooked. Today, farmers prefer a predictable hound that waits to be rewarded with a bone, to a battle with a 200-pound sow that rewards herself with a truffle.

But pigs have beaten dogs in their own territories, as this ingenious sow demonstrated in the field of hunting:

"Slut," as this animal was called, was very fond of sport, and would frequently walk a distance of seven miles in hopes of finding someone who was going out with a gun. She would point at every kind of game, with the curious exception of the hare, which she never seemed to notice. Although she would willingly back the dogs, they were very jealous of her presence and refused to do their duty when she happened to be the discoverer of any game, so that she was seldom taken out together with dogs but was employed as a solitary pointer. So sensitive was her nose that she would frequently point a bird at a distance of forty yards; and if it rose and flew away, she would walk to the place from which it had taken wing and put her nose on the very spot where it had been sitting. If, however, the bird only ran on, she would slowly follow it up by the scent, and when it came to a stop, she would again halt and point toward it. She was employed in the capacity of pointer for several years, but was at last killed because she had become a dangerous neighbor to the sheep.

from *Our Dumb Neighbors, or Conversations of a Father with his Children on Domestic and Other Animals,* Thomas Jackson (late nineteenth century)

A servile sow at the hunt is reassuring, but beware the wild boar, master of the unpredictable chase.

A Singular of Boars

Sus *Scrofa*, commonly known as the wild boar, is considered the most dangerous and clever of man's adversaries. Brave, fierce, equipped with a narrow body, long legs, and over 300 pounds of flesh, the boar is built for fighting. Its thick, bristly skin enables it to plunge into the prickliest of thickets, its bright, beady eyes and keen sense of smell and hearing alert it to approaching dangers, and its razor-sharp tusks can rip open an opponent with a toss of the head.

Boars live in forest lairs and venture out at dusk to rootle for their food — beetroot, potatoes, and Jerusalem artichokes being among the favorites. They

are sociable beings, living for the most part in groups, and following a rather ordinary existence of mud wallows, evening rootles, gruntings, farrowing, sex squabbles, and the cyclical shedding of winter coats for summer bristles. For all the boar's aggression and machismo, an old sow is usually head of the herd, preserving matriarchal tradition, while aged boars gracefully retreat into solitary retirement.

Wild boars appear in most parts of the world, scattered from Europe to Morocco and — further away — from Japan to Colombia. Their numbers drastically depleted in Europe during the Second World War but have since risen again, and the traditional boar hunt continues to be a popular sport. Nowadays, the risk of danger to man has been greatly reduced since the introduction of sophisticated weapons. Lost forever is the exaltation of the chase. Gone are the days when man, the hunter, entered the bush armed with just a sword or spear and met nature's most noble beast in man-to-pig combat.

The Tradition of the Boar Hunt

The Toast of the Master Hunt

Pledge me next the glorious chase
When the mighty boars ahead,
He, the noblest of the race,
In the mountain jungle bred
Swifter than the slender deer,
Bounding over Deccan's plain
Who can stay his proud career,
Who can hope his tusks to gain?

from *Sport in Many Lands*,
1877

Frescoes, paintings, murals, mosaics, and vase decorations depict boar hunting in Assyria, Egypt, and Greece as far back as 1300 B.C. The earliest hunters fought with bows and spears and then graduated to swords, plunging into the thickets on foot, on horseback, or behind a chariot. The ancient Greeks carried *machairas*, and considered the boar hunt one of the greatest mental and physical exercises in courage and discipline, an activity that prepared a man for eventual battle against his own kind. The boar hunt became an esthetic endeavor — an art.

Confrontations with boars feature heavily in Egyptian and Greek mythology. Several Greek gods actually took the guise of a boar to avenge themselves of an enemy. Apollo was one of these. (He gored Adonis to death for insulting Aphrodite.) Set was another. He chose to castrate his unfortunate enemy, Osiris. Many ancient heroes and gods ended their days on a tusk. Tammuz, Cretan, and Zeus were all savaged in thickets, and Ancaseus of Arcadia, the helmsman of the Argonauts, was killed by a wild boar who was trampling down his precious vineyard. Other heroes, who survived their boar hunts, went on quests or labors, and accounts of their adventures have been passed down through generations.

A Famous Hero

Ulysses, the most renowned of the Greek heroes, proved himself a worthy warrior at an early age, living up to his name, which in Greek means "man of wrath." He was given the name by his grandfather, Autolycus, who invited the boy to stay with him on Mount Parnassus. Soon after his arrival, Ulysses was taken on his first boar hunt. As he was already the fastest runner in Greece, Ulysses sprinted ahead of the other pursuers and tracked the boar with his famous hound, Argos:

He came on a great boar lying in a tangled thicket of boughs and bracken, a dark place where the sun never shone, nor could the rain pierce through. Then the noise of the men's shouts and the barking of the dogs awakened the boar, and up he sprang, bristling all over his back, and with fire shining from his eyes. In rushed Ulysses first of all, with his spear raised to strike, but the boar was too quick for him, and ran in, and drove the sharp tusk sideways, ripping up the thigh of Ulysses. But the boar's tusk missed the bone, and Ulysses sent his sharp spear into the beast's right shoulder, and the spear went clean through, and the boar fell dead, with a loud cry. The uncles of Ulysses bound up his wound carefully, and sang a magical song over it, as the French soldiers wanted to do to Joan of Arc when the arrow pierced her shoulder at the siege of Orleans. Then the blood ceased to flow, and soon Ulysses was quite healed of his wound. They thought that he would be a good warrior, and gave him splendid presents, and when he went home again, he told all that had happened to his father and mother, and his nurse Eurycleia. But there was always a long white mark or scar above his left knee, and about that scar we shall hear again, many years afterward.

from *Tales of Troy and Greece*, edited by Andrew Lang

Some Notorious Bristlies

The Erymanthian Boar was the fourth labor imposed on Hercules by King Eurystheus. The beast had earned itself such a savage reputation that the King, in fear and trembling, hid in a jar until it was captured. The Erymanthian Boar had to be caught alive and presented to the King, and it led Hercules a song and dance up and down the slopes of Mount Erymanthis but was finally outwitted and cornered in a snowdrift. Hercules carried it on his shoulders to Mycenae, where he deposited it in the market place, only to find later that it had quietly trotted back to its mountain slope. A sheepish King Eurystheus emerged slowly from his jar.

The Crommyum Sow was an enormous, vicious creature that uprooted crops far and wide and generally caused a great deal of havoc. It was eventually killed by Theseus, who from then on ran into pig problems wherever he went. On another labor, he had to wrestle with a charming character called Cercyon the Arcadian, who was notorious for crushing his victims to death. Cercyon's crushing mania may have had something to do with his father, Branchus, being named after grunting pigs; Cercyon himself was associated with a pig cult.

The Calydonian Boar was the most famous of all the bristlies. Sent by Artemis to ravage the lands, the Calydonian Boar brought out the Greek heroes in droves: Telamon, Hyleus, Theseus (again), Jason, Nestor, Iphicles, Ancaeus, Eurytion, Amphiaraus. Meleager organized the hunt and invited along his ladylove, Atalanta, to share in this special treat. By all accounts, the boar had a

field day. Nestor was treed, Ancaeus castrated and disemboweled, Telamon tripped over a tree stump, and Hyleus killed Eurytion by mistake. Jason flung javelins far and wide, and Iphicles managed to graze the animal, but it was Atalanta who finally speared it. Meleager finished off the job and presented Atalanta with the prized pelt. This gesture did not please Meleager's uncles, however, who were jealous of Atalanta's victory. They took the hide from her, and in a fit of temper Meleager killed them. This, in turn, displeased Meleager's mother, Althaea, who avenged her brothers by causing Meleager's death.

Ceremony and Pageantry

By the Middle Ages, the boar hunt had become a spectacular spring ritual of both village and court life. Special huntsmen tracked down the boar and marked up an area designated for the hunt; then the dogs were sent in to flush the animal out of the undergrowth. The boar, ever heroic, charged, putting up a bitter fight, often killing dogs and wounding horses and men. Eventually, the snapping mongrels cornered it, while a mounted huntsman thrust at it from the saddle with an *estoc* or pointed sword:

This was a dangerous procedure. One thrust of the fine point of such a sword would neither kill nor stop its onslaught, and the rider could be easily toppled from his saddle. The great four-teenth-century writer on hunting, Count Gaston de Foix, otherwise [known as] Gaston Phoebus, con-sidered that to kill a boar in this fashion, when the animal was not held by mastiffs, was the greatest feat of all and a "fairer thing and more noble" than killing with a spear.

from *Hunting Weapons*,
Howard L. Blackmore

Taking up the chase on foot must have been considered more coura-geous still. Gaston Phoebus's advice to horseless hunters is clearly not directed at the faint-hearted:

Hold your spear about the middle, not too far forward, lest he strike you with his tusks, and as soon as the point has en-tered the body, take the haft of the spear under your armpit, and press and push as hard as you can and never let go the haft, and if the beast be stronger than you, then you must turn from side to side as best you can without letting go the haft, until God comes to your aid or other assistance reaches you.

Edward II, another four-teenth-century enthusiast of the sport, obviously shared Phoebus's respect for this most formidable quarry:

Nor is there any beast that he could not slay sooner than that other beast could slay him? Be they lion or leopard — and there is neither lion nor leopard that slayeth a man at one stroke as a boar doth, for they mostly kill with the raising of their claws and through biting, but the wild boar slayeth a man at one stroke as with a knife.

Other notable blue-blooded pursuers of pig were James I, the Emperor Maximilian I, Philip IV of Spain, and that zealous participant of all lusty sports — Henry VIII:

Henry VIII was a great lover of the boar hunt and, like all princes, enjoyed the privilege of possessing weapons of fine quality. The 1547 inventory of his arms and armor included at the Tower of London:

Bore speares wt asshen staves trymed wt crymesyn velvet and fringed wt redde silke.
Bore speares knotted and lethered.
Bore speares wt asshen staves trymed wt lether.
Bore speares graven and gilte.

From Noble Sport to Sorry Spectacle

At various times throughout history, men become bored with heroics and turn to spectacles for entertainment. The Romans, for instance, grew weary of the difficult chase and — unlike the Greeks, who had respected the power of the boar and the challenge of the hunt — preferred to sit in comfort and safety and watch wild boars being hurled into an arena and massacred for fun. And in Europe, by the late seventeenth century, the pageantry of the boar hunt as portrayed by Gaston Phoebus was reduced to a theatrical enterprise, where the hapless animal did not stand a chance and was butchered publicly to satisfy jaded appetites. A painting by Jakob Philipp Hackert in the Museo di Capodimonte depicts a boar hunt of Ferdinand IV of Naples in 1785, and is described by Howard L. Blackmore in *Hunting Weapons* as follows:

The hunt took place at Cassano, where the boars were driven by dogs over marked-out courses across a level plain. The mounted huntsmen pursued them with short light lances, which could be plunged or thrown when the boar was in range. But the absence of cover, the flagged runs, the groups of servants controlling the dogs, and the long line of carefully marshaled spectators show to what artificial levels the hunt has degenerated.

The eighteen and nineteen hundreds saw the introduction of firearms, and the risks to man were further lessened. The former close contact between hunter and hunted — the thrill of adversaries meeting on more or less equal footing — was no more.

An acquaintance of ours, who before the Second World War was an old hand at boar hunting in Poland, recalls:

Poland had thousands of boars, especially in the forests and mountains. In fact, the Carpathians were so thick with boars, you could hear them for miles around, crashing and foraging through the undergrowth. The best time for the hunt was in winter, when you could follow their tracks. A good tracker would get up in the morning, inspect the tracks, and be able to tell you exactly where the boars were sleeping and how many there were. Then the beaters were sent round behind the boars to drive them out of the forest toward the hunters.

Contrary to common belief, a boar will never attack unless provoked or wounded, and then woe betide anyone in its path. The same goes for a mother and her young; she will defend them to her death. The boar is also tremendously resilient. I've seen bullets bounce off its skin, it's so thick. Once in full gallop, a boar is as quick as lightning, and it would be very foolish to try and shoot it head-on. Even dying boars can kill a man. I remember one that, although unconscious, managed to struggle to its feet and wound someone before it collapsed again. The smaller boars are actually more dangerous, being swifter and fiercer than the larger ones.

We sometimes hunted with dogs — mongrels and fox terriers were the best. I remember one time we were hunting an old boar that had evaded us on several occasions and knew it was wanted. Its instincts for survival were really aroused, and it became very crafty, leading us through thick undergrowth and then disappearing and lying low. At one point, my friend, thinking he had wounded it, rushed into the wood with his gun, only to be ambushed by the boar, which was hiding. It sprang out of the thickets on top of him. He lost his gun and received a huge slash on his leg and would probably have been savaged to death had not the dogs come to his rescue by causing a diversion. Of course, that was an unusual accident. The hunt is not nearly as risky as it used to be when they used spears and swords.

W. Skarbek-Borowski

Pig-Sticking in India

In India, at the beginning of the nineteenth century, boar spearing or pig-sticking became the favorite sport of the British regiments. This was the grand era of the British Empire, of maharajahs, Sahibs, Kipling, polo, and tiffin. Pig-sticking was more than just a pastime; it was a passion. It was also extremely dangerous — certainly much more so than contemporary boar hunts.

On horseback and armed only with a light bamboo spear, a soldier needed speed, agility, and luck against his small but ferocious opponent.

Major A. E. Wardrop, in his book *Modern Pig-Sticking* (published in 1914), has some advice for novices:

1. Always gallop fast at a pig: never walk.
2. Always ride at, or take a charging pig at an angle, not end-on.
3. Spear well forward.
4. Have a sharp spear, and do not hold on to it.
5. A horse can go where a pig goes.
6. If you lose a pig, cast for'ard.
7. You must lie first in a run.
8. Always run with a loose rein.

Brigadier General Gaussen, a famous pig-sticker and another British officer stationed in India at the beginning of this century, survived some hair-raising escapades in his adventurous life. On this occasion, he recounts how he went pig-sticking in pyjamas:

During the hot weather in Allahabad, there was a bad famine, which brought pigs into the cantonments at night. I had had a very late and thirsty night at some regimental show and went to bed in the open in our mess compound. It seemed I had barely got to sleep, when I was awakened by a great noise. As a matter of fact, it was dawn, and I gradually realised that a lot of servants were shouting out that there was a large boar in the compound. By the time I collected my wits, a groom had saddled a pony, who was picketed in the open close by, and another had brought out a spear. I got into the saddle as I was, for true enough, there in the far corner, a big boar was trying to get over the high mud wall. As I started off, he managed to get over, and I raced for the gate in time to see him cross the road and get into the next compound. As the wall was low, I popped over it and nearly got hung up in an array of mosquito netting and beds. The pig swung left and made for the river, but when I got up to the bank, not a sign of him could I see, and yet it was all open sand. Then I spotted a small mud hut made for a guardian of a melon patch, and I knew the pig must be inside. So I cautiously approached the entrance, and sure enough, with a Woof! Woof! out charged the boar and gave me a good opening. When I rode back feeling rather elated, I was met by an officer with a long face. It appears that a newly married Sapper, who lived opposite, had been over to the C.O., who luckily was not up, accusing me of riding "all but naked" into his compound and terrifying his wife. When he heard that I was after a pig, all was well, but no doubt I did look a bit mad charging into their beds in pyjamas and a flapping jacket!

from *The Illustrated Sporting and Dramatic News*, December 23, 1933

The exaltation of the hunt was usually celebrated through song and drink in the officers' mess. Between countless glasses of port, the pig-stickers raised the roof with rousing ballads such as "The Nagpur Hunt" (to be sung to the tune of the hymn, "Oh God our Help in Ages Past"):

Mohurrum Meet, 1908

A band of pig-stickers set out
  From Kamptee and Nagpur,
All bound for the Mohurrum Meet,
  To hunt the Mighty Boar.

The Captain bold was Gibson, he
  Called Pat, and Mrs. G.
And Hoskyns and young Jones they made
  An eager company.

At Tooljapur young Coventry
  Appeared upon the scene,
And walked us off to dinner all,
  A merry crowd I ween.

Next morning Pratt arrived agog
  To slay, and heartily
We welcomed him a brother bold
  For any devilry.

And now wake up my Pegasus,
  Come put your best leg forth,
Record in your most fluent verse,
  And go for all your worth.

The best day in the annals of
  Our famous hunt must now
Be handed down by you to all,
  "A sorry steed I trow."

With spavins, ringbone, splints, and strains,
  And windgalls not a few;
You know you simply have to go,
  You miserable screw.

Before we started from the camp
  A panting beater came,
Reporting that a mighty boar
  Was lying near; one game

To kill his man; so mounting quick
   We rode to lay him low.
Roused from his lair in rage he rushed
   Upon his nearest foe.

Mahomed Ali just in time
   To save his pony speared
The head; then all with wild hooroosh
   Joined in the fray and cheered.

After a jink or two young Jones,
   Mounted on "Foccus Dog"
(That hardy, game old veteran),
   Got on. It seemed the hog

Was surely his without a doubt,
   When slipped his saddle round,
And J was clean out of the hunt,
   And nearly kissed the ground.

Then Gibson with unerring spear
   Transfixed the boar's thick hide,
In less time than it takes to write
   That old manslayer died.

We gazed on his great bulk with awe.
   By jove! he was a snorter,
The measure tape proved him to be
   Five-thirty and a quarter.

No time was there to mourn for him,
   The work was yet to do,
The soor-log were many still,
   The spears were very few.

Then Blenkie, Jones, and Coventry,
   With Hirst (a new recruit),
Pursued a swiftly running hog,
   Caught up and killed the brute.

But not before his mark he left
   On Hirst's game little Tat.
He cut it badly in the hock,
   'Twas of the genus "rat,"

But very game. The Gibson crew
   Found and rode another
Stout member of the porcine breed,
   To No. 2 a brother.

Brave Mrs. G on trappy ground
   Got on him first, and G
Followed in dread, being sure that now
   A widower he'd be.

She gave a real good spear, and then
   That gallant pig he fought
At all and sundry charging in,
   And set his foes at naught.

But finally he, too, was slain.
   Young Hoskyns' turn came next
After a weighty gentleman
   Who ran until perplexed,

And out of breath with running hard
   Sought refuge in the river,
After the Hoskyns' spear had pierced
   The region of his liver.

Now up and down on foot they fought
   Close by the river flood,
Until this gallant boar at last
   Fell lifeless in the mud.

"To horse! To horse!" the Captain cried,
   "No place for sluggards here,"
And out we went to beat the Ban,
   Or did we beat a Bir?

I quite forget, but anyhow,
   A goodly boar soon broke,
And off we rode to lay him out,
   Each larruping his moke.

Save Gibson none had ere a chance,
   And his was second one.
So died a likely pig before
   The fight had well begun.

Another boar reported was,
   Hopes of the bag enhancing
Were great. We went to turn him out
   When lo on us advancing

We saw him come, formed line, and met
   The foe, but turning tail
This coward pig fled swiftly off.
   Flight was of no avail.

J took the first ignoble spear,
  Soon died the coward boar;
His weight was near 300 lbs.,
  His height was thirty-four.

Six boars were slain; another one
  Claimed our attention still;
And Mrs. G, on Cigarette,
  Rode on his tail until

He charged, and charging met the spear,
  A good one. In each hand
Grasping a lance, G got him down
  And quickly made an end.

So seven boars by seven spears
  Were slain: can one do better?
This day should be recorded in
  Our annals as "red letter."

Divided up the inches of
  Those seven pig when done
Gave thirty-three to each; their weight
  Panned out 261.

All honor to our Captain bold,
  All honor to his crew,
For showing us such ripping sport,
  Jowahir Singh, Paikoo;

Abdullah with its faithful oont,
  The Buddoo, don't forget.
And now to dodge the eggs and cats
  I'll not remain, you bet.

F.W.C.J.

Lesser Sports

Another human pastime, which is considered fairly harmless by men but which, nevertheless, must be a humiliating experience for pigs, is "pig-running." In this sport, a pack of screaming children pursues a piglet. They invariably capture the frightened animal by throwing themselves on top of it. The piglet squeals with fright, while the predominantly adult spectators roar with delight. A more equitable variation of "pig running" is "pig greasing." The general idea is the same except that the piglet is better oiled and at least stands some chance of escape.

From the glory of the ancient boar hunts to the ignominy of terrorizing a greased piglet — how strange and futile are the pursuits of man!

Pig Myth

Let us make man in our image, after our likeness; and let them have dominion over the fish of the sea, and over the birds of the air, and over the cattle, and over all the earth, and over every creeping thing that creeps upon the earth.

Genesis 1:26

With these wise words ringing in our ears, we have succeeded in subduing the animal kingdom. We have managed to destroy the natural habitat of most species, to slaughter the proudest and fiercest in sport, and to snatch away the dignity and spirit of the untamed by parading them in circuses and caging them in zoos. Physical contact and understanding between human beings and untamed animals has been generally vicarious — most of us are content today to remain in our armchairs, watching (and believing in) films of egregious monsters besieging innocent people. We need to see these animals made to submit to our "superiority," while we fondle our household pets. We are capable of loving only those whom we can control.

The pig that most of us readily relate to today is the quiet nesting type, happily ensconced on a pile of hay. But, as the early boar hunts demonstrated, man was once capable of feeling both an affinity with and a respect for the power of the pig. In the past, a much more sophisticated and logical attitude was taken toward pigs. They were an integral part of the general order of things, and their sexual differences and characteristics were incorporated into the mythology of specific cultures.

The Serenity of the Sow

The sow, a picture of serenity and comfort, was frequently considered the epitome of all that is pure and good. Her maternal instincts were thought to protect, and her intuition was revered. In myth and legend, she was usually portrayed as white — the color of freshness, strength, and fertility — and was sometimes connected with mother earth deities.

Tellus is the Roman goddess of the earth, from which seeds grow and which receives the dead. Never failing, she gives nourishment to all, and when the soul leaves the body, man returns to her, and not only men, but all living things, for she is both the mother (terra mater) and the grave of everything that lives, and as some have believed, the first of all gods. Tellus receives the seeds into herself and causes them to grow. Therefore, she was invoked together with Ceres at the festival when the harvest was brought in. She also had her own feast day, the Fordicidia, on 15 April, when pregnant pigs were sacrificed to her since the time of King Numa, for such a sacrifice, it was held, made the earth more fertile.

Sabine G. Oswalt

The suckling sow, as a symbol of fertility, was a sign of wealth and security as well. When the god of the River Tiber told Aeneas that a sow would mark the site of Rome, he was providing the Trojan hero with an image of the grandeur and stability of the future city:

And that this nightly vision may not seem
Th'effect of fancy, or an idle dream,
A sow beneath an oak shall lie along,
All white herself, and white her thirty young,
When thirty rolling years have run their race,
Thy son Ascanius, on this empty space,
Shall build a royal town, of lasting fame,
Which from this omen shall receive the name.

from The *Aeneid*, Virgil
(translated by John Dryden)

The sow and her young were later celebrated in Christian times, usually as decorative images in the windows and other parts of churches. The white sow, like the Virgin Mary, possibly signified the eternal mother — both exemplify the Christian maternal ideal of humility and purity — but it is more likely that the sow was associated with the bounty of the thanksgiving festival, when God was praised for his goodness in bestowing a healthy harvest.

The sow's fertility was also tied to themes of light and springtime. Demeter, the Greek goddess of harvest and seedtime and mother of Persephone, headed a white sow cult led by a magician swineherd called Eubuleus, and because she possessed similar attributes to the sow, she assumed the anthropomorphic form. When Persephone was abducted by Pluto and forced to marry him in Hades, Demeter roamed the earth unceasingly in search of her. The earth turned into a wasteland until she found her daughter, at which time it became fruitful again.

Persephone represents the growth and harvesting of the corn. She goes down to the underworld every year in summer, when the harvest has been brought in, before the summer heat. During the heat, the fields lie dry and barren. She returns again in the autumn, when the fields are ploughed and the seed is sown. The first green appears later in autumn and continues through the months, for winters in Greece are mild, and during all this time Persephone lives with Demeter above ground.

Sabine G. Oswalt

The Power of the Boar

The boar in symbolism has rarely transcended his reputation as a fierce fighter of the field. While he does appear in many guises, he is mostly associated with power and male supremacy. Both monarchs and gods have esteemed his strength. King Richard III of England chose a white boar as his emblem and so declared, in heraldic terms, his belief that the boar was more honorable and manly than any other beast. And in Indian legend, the god Vishnu took the form of the Boar Avatar and killed the Demon Hiranyaksha, who was holding a flooded earth captive. The Boar Avatar darted across the sky, plunged into the water and, using his powerful snout, traced the earth by its smell. He then raised the planet back to the surface with his strong tusks. In some stories, Vishnu is depicted as a giant with a boar's head, carrying in his arms the goddess Earth.

To the ancient Greeks, the wild boar was not only a symbol of strength, but also of turmoil and upheaval. He was sacred to Ares, the god of war, who himself portrayed boar-like tendencies in battle, being angry, unpredictable, and headstrong:

For blind, oh, blind, women, is he that perceived not that Ares in the form of a boar sets all evils in commotion

from an ancient fragment preserved by Plutarch

Just as the sow represented light, the boar often signified darkness. The Egyptians were permitted to eat boars' flesh only during midwinter, the dark season. Curiously, though, the warrior Vikings connected the boar with the light, nourishing qualities usually ascribed to the sow. They worshipped a god called Frey, who ruled over crops, trees, rain, and sunshine, and who rode across the sky from east to west in a chariot drawn by a golden boar. Frey was honored by a feast where, fittingly enough, roasted boar's head was served.

The boar was also considered a *figure* of darkness, a usurper, robbing the true and anointed king, the ruler of light, of his throne:

The wretched, bloody, and usurping boar,
That spoil'd your summer fields and fruitful vines,
Swills your warm blood like wash, and makes his trough
In your embowell'd bosoms — this foul swine
Is now even in the center of this isle.

King Richard the Third,
Act V, Scene II,
Shakespeare

The Celts greatly feared the dark power of the boar. Even the bull, a more common symbol in their culture, was not thought to be so terrifying. Celtic burial grounds have been discovered wherein the remains of warriors are accompanied by parts of a boar's anatomy (usually the strong tusks), reflecting a belief in the boar either as a fierce fighter or an evil influence. In an old Irish folktale, "The Death of Dermott," The Boar of Ben Gulban is known to carry a curse:

"They hunt the Boar of Ben Gulban," Finn told him. "They are foolish to do that. Already the boar has wounded a score of the beaters."

"I heard a hound baying in the night," said Dermott, "and that was what brought me here."

"Leave the hill to the boar and his hunters, O'Duivna," said Finn. "I do not want you to be here."

"Why should I leave the hill?" Dermott answered. "I have no dread of a boar."

"You should have," said Finn "for you are under prohibitions about hunting a boar."

"My father was cursed on account of a boar," said Dermott, "I know that."

"The curse was to fall on your father's son and on no one else, O'Duivna," Finn said. "It has been told you by me, remember."

"Even so," said Dermott, "it would be craven of me to leave the hill before I had sight of the boar. Here I stay. But would you leave your hound with mine?"

Finn did not answer. He went down the side of the hill, and his hound followed him, leaving Dermott a solitary man with the hound Mac an Cuill beside him. There was a tearing noise, and Dermott knew that the Boar of Ben Gulban was coming up the hill, and he knew that none was in pursuit of him because there was no baying of hounds behind him. No men, no hounds were there to hem him round.

Dermott unslipped Mac an Cuil. But the hound cowered before the bristled, gnashing brute.

He slipped his finger into the string of his lesser spear, the Gae Bwee, and made a careful cast. The spear struck the boar between his little eyes. But not a single bristle was cut, not a gash or scratch was made upon him.

"Woe to him who heeds not the counsel of a good wife," said Dermott O'Duivna to himself.

He drew the Begalltach from his belt and struck the wild boar. But even then, the boar pitched him so that he fell along the bristly back. Then the boar turned and crashed down the side of the hill. He rushed back and reached the summit again. There he pitched Dermott down and gored and ripped him. And Dermott lay on the ground, not able to raise himself, writhing with the pain of his deep, wide wounds.

On the track of the boar came the hunters: Finn and four chiefs of the Fian with them. They found Dermott where he was lying; and Oisin and Caelte raised him up.

"What grief to see a hero torn by a pig," said Oisin. "What grief to us to see Dermott O'Duivna in this plight."

"I grieve," said Finn, "that the women of Ireland are not here to gaze on him and to see the beauty and grace they found in him spoiled by the pig's gashes."

Despite their dread of the boar, the Celts, like the Vikings, immersed themselves in bestial rites at festivals held to pay tribute to the sun and the changing seasons. These celebrations took place during the spring and autumn equinoxes (March 21 and September 21) and the summer and winter solstices (June 21 and December 21). The Druids donned animal skins to represent horned gods. Huge hill-top fires were lit for the occasion, which illuminated one horrified Christian spectator:

Some dress themselves in the skins of herd animals; others put on the heads of horned beasts; swelling and madly exulting if only they can so completely metamorphose themselves into the animal kind that they seem to have completely abandoned the human shape.

St. Caesarius of Arles
(seventh century)

It was not long before such "paganism" was suppressed by the Christian church, and the fires of the winter solstice were transformed to candlelight ceremonies and yule logs honoring Christ's birth; however, the boar's head remained part of the celebratory feast.

Boar's Head Carol

Traditional

A legend well over 500 years old tells us that a student at Queen's College, Oxford, was taking a lonely walk one Christmas morning on a hill near Oxford town when he was attacked by a hungry boar. Otherwise weaponless, the student thrust his copy of a book by Aristotle (legend does not say which one) down the beast's throat, choking him to death. That night the dons and students of the college ate the boar. Every Christmas since then, continues the legend, a boar's head is carried into the hall on a platter, formally placed on the high table, while the carol is sung. The hill where the scholar put Aristotle to valiant use is still called Boar's Hill.

The carol is printed in 1521 by Jan van Wynken de Worde in *Christmasse Carolles*, probably the first volume ever printed in England to contain music.

Here are the meanings of the Latin parts of the macaronic verses:

Quot estis in convivio — as many as are dining together

servire cantico — to serve with a song

In reginensi atrio — in the hall of the Queen

Caput apri defero, reddens laudes Domino — I bring in the boar's head, singing the praises of the Lord.

Queen's College, Oxford, version

Traditional (M.S.)

1. The boar's head in hand bear I, Be-decked with bays and rose-ma-ry; And I pray you, my mas-ters, be mer-ry, Quot es-tis in con-vi-vi-o:

CHORUS
Ca-put a-pri de-fe-ro, Red-dens lau-des Do-mi-no.
End here.

2. The boar's head, as I un-der-stand, Is the rar-est dish in all this land, Which thus bedecked with a gay gar-land, Let us ser-vi-re can-ti-co:
Repeat CHORUS

3. Our stew-ard hath pro-vi-ded this, In hon-our of the King of bliss, Which on this day to be ser-ved is, In Re-gi-nen-si a-tri-o:
Repeat CHORUS

from *A Treasury of Christmas Songs and Carols*,
Henry W. Wilson

The Beast Within Us

It is hard to determine exactly why pigs were not always sanctified, why they frequently became associated with all that is abhorrent in human nature.

Maybe the answer has something to do with the tenuous line between love and hate. In any love affair, our nearest and dearest often become victims of our darker side. That famous sorceress Circe, for example, tormented Ulysses by changing his men into the animals of which they reminded her.

lot of Eurylochus leaped out, and, weeping for fear, he led his twenty-two men away into the forest. Ulysses and the other twenty-two waited, and, when Eurylochus came back alone, he was weeping, and unable to speak for sorrow. At last he told his story: they had come to the beautiful house of Circe, within the wood, and tame wolves and lions were walking about in front of the house. They wagged their tails, and jumped up, like friendly dogs, round the men of Ulysses, who stood in the gateway and heard Circe singing in a sweet voice, as she went up and down before the loom at which she was weaving. Then one of the men of Ulysses called to her, and she came out, a beautiful lady in white robes covered with jewels of gold. She opened the doors and bade them come in, but Eurylochus hid himself and watched, and saw Circe and her maidens mix honey and wine for the men, and bid them sit down on chairs at tables, but, when they had drunk of her cup, she touched them with her wand. They they were all changed into swine, and Circe drove them out and shut them up in the sties.

Next morning, Ulysses divided his men into two companies, Eurylochus led one company and he himself the other. Then they put two marked pieces of wood, one for Eurylochus, one for Ulysses, in a helmet, to decide who should go to the house in the wood. They shook the helmet, and the

When Ulysses heard that, he slung his sword-belt round his shoulders, seized his bow, and bade Eurylochus come back with him to the house of Circe; but Eurylochus was afraid. Alone went Ulysses through the woods, and in a dell he met a most beautiful young man, who took his hand and said, "Unhappy one! how shalt thou free thy friends from so great an enchantress?" Then the young man plucked a plant from the ground; the flower was as white as milk, but the root was black; it is a plant that men may not dig up, but to the Gods all things are easy, and the young man was the cunning God Hermes, whom Autolycus, the grandfather of Ulysses, used to worship. "Take this herb of grace," he said, "and when Circe has made thee drink of the cup of her enchantments the herb will so work that they shall have no power over thee. Then draw thy sword, and rush at her, and make her swear that she will not harm thee with her magic."

Then Hermes departed, and Ulysses went to the house of Circe, and she asked him to enter, and seated him on a chair, and gave him the enchanted cup to drink, and then smote him with her wand and bade him go to the sties of the swine. But Ulysses drew his sword, and Circe, with a great cry, fell at his feet, saying, "Who art thou on whom the cup has no power? Truly thou art Ulysses of Ithaca, for the God Hermes has told me that he should come to my island on his way from Troy. Come now, fear not; let us be friends!"

Then the maidens of Circe came to them, fairy damsels of the wells and woods and rivers. They threw covers of purple silk over the chairs, and on the silver tables they placed golden baskets, and mixed wine in a silver bowl, and heated water, and bathed Ulysses in a polished bath, and clothed him in new raiment, and led him to the table and bade him eat and drink. But he sat silent, neither eating nor drinking, in sorrow for his company, till Circe called them out from the sties and disenchanted them. Glad they were to see Ulysses, and they embraced him, and wept for joy.

from *Tales of Troy and Greece*, edited by Andrew Lang

Because pigs have been used as objects of our abuse, they have carried the responsibility for a number of our sins, notably sloth, greed, and lust. This state of affairs took on an absurd cast in Europe during the late Middle Ages, when it was felt that pigs were not just corrupt, but downright cunning:

Thus, in 1456, a pig was sentenced to be burned in Rhine for having killed and eaten a small child. Another pig was convicted on a similar charge in Amsens in 1463. In such instances, the animals were sometimes tortured — in order, according to one view, that their grunts and squeals could be interpreted as confessions....

The anthropomorphic character of animal trials was sometimes emphasized by dressing the animal in clothes and tying the creature in a sitting position during the proceedings. In 1386, a pig — tried in Falaise, Normandy, for damaging the face and arm of a child — was dressed in clothes and sentenced to be maimed as the child had been maimed. In the Church of the Holy Trinity in Falaise, there was a fresco depicting the trial, but it was painted over in 1820 by order of the Church authorities.

from *The Witchcraft World*,
G. L. Simons

Perhaps nowhere is the victimization of the pig — and the implications of that victimization — so vividly realized as in William Golding's *Lord of the Flies*, a moral fable about what can happen when we are unable to escape from the forces of evil — the original sin or the "beast" — that lurk within all of us.

A group of young boys, who are being evacuated during an atomic war, crash-land on an island. Dissension over leadership quickly springs up among the boys, who are all very different in character — Ralph is the most popular, Jack is a coarse bully, Simon is artistic and spiritual, and Piggy is a middle-of-the-roader. Golding's message begins to emerge: for society to survive, it must have some kind of structure based on reason and the maintenance of law and order. Man lives by his basic instincts, and anarchy results when he is allowed to return to the primitive savage stage. The boys' world on the island becomes a parody of the grownup world at war.

The title, "Lord of the

Flies," refers to the devil, Beelzebub. In Golding's story, Beelzebub or the "beast" is symbolized by a pig's skull grinning on a stick and surrounded by swarming flies. Pigs continue to represent hunted and sacrificial victims through-out the book.

The boys first discover the conflict between instinct and culture when they come across a wild pig in the jungle. Although they want to kill it, they can't bring themselves to do so. Instead, they relieve their frustrations by playing games. One of the boys pretends to be a pig:

"Kill him! Kill him."

All at once, Robert was screaming and struggling with the strength of frenzy. Jack had him by the hair and was bran-dishing his knife. Behind him was Roger, fighting to get close. The chant rose ritually, as the last mo-ment of a dance or a hunt.

Kill the pig! Cut his throat! Kill the pig! Bash him in!"

Ralph too was fighting to get near, to get a handful of that brown vulnerable flesh. The de-sire to squeeze and hurt was overmastering.

Jack's arm came down; the heaving circle cheered and made pig-dying noises. Then they lay quiet, panting, listen-ing to Robert's frightened snivels. . . .

"That was a good game."

"You need a real pig," said Robert, still caressing his rump, "because you've got to kill him."

The games soon become real. The boys, led by Jack, begin hunting and killing wild pigs for pleasure. In one harrowing scene, a sow and her young are torn to pieces. Pig hunting is now a ritual to ease tensions within the group and to satiate primitive desires. Chanting and dancing reach frenzied proportions with cries of "Kill the pig. Cut his throat. Bash him in." Then killing pigs is suddenly no longer enough — the boys sadistically turn on each other. Simon is the first to recognize the "beast," and is killed because he doesn't co-operate. The chant takes on a more sinister ring: "Kill the beast! Cut his throat! Spill his blood!" Finally Ralph ends up a "pig." The gang pursues him through the jungle, and he is saved just in time by the arrival of a British naval officer who has landed on the beach.

Piggy is the most interesting character in the book. As his nickname suggests, Piggy is fat. He is also asthmatic and wears glasses, without which he is virtually blind. He is ponderous and plodding, and the other boys either hate him or ridicule him. Although he recognizes the "beast," he lacks the charisma to persuade his companions to remain civilized. Piggy's greatest mo-

ment is at the time of his death, when he challenges Jack and the gang:

A great clamor rose among the savages. Piggy shouted again.

"Which is better — to have rules and agree, or to hunt and kill?"

Again the clamor and again — "Zup!"

Ralph shouted against the noise.

"Which is better, law and rescue, or hunting and breaking things up?"

Now Jack was yelling too, and Ralph could no longer make himself heard. Jack had backed right up against the tribe, and they were a solid mass of menace that bristled with spears. The intention of a charge was

forming among them; they were working up to it, and the neck would be swept clear. Ralph stood facing them, a little to one side, his spear ready. By him stood Piggy, still holding out the talisman, the fragile, shining beauty of the shell. The storm of sound beat at them, an incantation of hatred. High overhead, Roger, with a sense of delirious abandonment, leaned all his weight on the lever...

...The rock struck Piggy a glancing blow from chin to knee: the conch exploded into a thousand white fragments and ceased to exist. Piggy, saying nothing, with no time for even a grunt, traveled through the air sideways from the rock, turning over as he went. The rock bounded twice and was lost in the forest. Piggy fell forty feet and landed on his back across that square, red rock in the sea. His head opened and stuff came out and turned red. Piggy's arms and legs twitched a bit, like a pig's after it has been killed. Then the sea breathed again in a long, slow sigh, the water boiled white, and pink, over the rock; and when it went, sucking back again, the body of Piggy was gone.

Bedeviled

Because of pigs' supposed inclination toward shamelessly sinful behavior, they commonly represent the devil or the devil's auxiliary in religious parables:

And Jesus asked him, "What is your name?" He replied, "My name is Legion; for we are many." And he begged him eagerly not to send them out of the country. Now a great herd of swine was feeding there on the hillside; and they begged him, "Send us to the swine, let us enter them." So he gave them leave. And the unclean spirits came out, and entered the swine; and the herd, numbering about two thousand, rushed down the steep bank into the sea, and were drowned in the sea.

Mark 5:9-13

Also, according to a medieval legend, St. Anthony Abbot, after a grueling war of nerves with the devil, emerged victorious, and his adversary was changed into a pig, which forever followed him around at his heels.

Are you going to (meat) me at the place named?

Pig Flesh — Sacred or Accursed?

There appears to be no happy medium with the pig: it is sanctified by some; damned by others. And so, certain taboos and rituals have grown up around the consumption of its flesh. Eating an animal that you either revile or revere (depending on your religious allegiance) is not an activity to be taken lightly. Yet pork taboos, be they prohibitive or hallowed, were probably founded less in divine theory than in a practical concern for cultural and economic stability.

And the swine, though he divide the hoof, and be clovenfooted, yet he cheweth not the cud; he *is* unclean to you.

Leviticus 11:7

The harmful parasites in undercooked pork, which can be transmitted to the eater, do in a sense render the animal "unclean;" however, the original reason for Jewish and Moslem taboos against pork may not have been just hygienically motivated:

The Bible and the Koran condemned the pig because pig farming was a threat to the integrity of the basic cultural and natural ecosystems of the Middle East.... Until their conquest of the Jordan Valley in Palestine, beginning in the thirteenth century B.C., the Hebrews were nomadic pastoralists, living almost entirely from herds of sheep, goats, and cattle. Like all pastoral peoples, they maintained close relationships with the sedentary farmers who held the oases and the great rivers.... Within the overall pattern of this mixed farming and pastoral complex, the divine prohibition against pork constituted a sound ecological strategy. The nomadic Israelites could not raise pigs in their arid habitats, while for the semi-sedentary and village farming populations, pigs were more of a threat than an asset.

from Cows, Pigs, Wars and Witches, Marvin Harris

The other side of the coin is represented by the Maring tribesmen of New Guinea's Bismark Mountains, who hold pigs as sacredly as do Christians the Eucharist. For these people, the eating of pork is a consecration. Every twelve years, in order to reduce the pig population — which by then has grown to such an extent that the natural resources of the area are threatened — a ritual festival is held. For almost twelve months, various tribes periodically go to war against each other, and numerous pigs are sacrificed before each battle. Once the adult pig population is decimated, the fighting stops, the tribal ancestors are appeased, and all efforts are spent raising a new generation of tuskers.

Peace reigns for another twelve years, and then the pig ritual begins again.

The Maring tribes' unique relationship with pigs... includes raising pigs to be members of the family, sleeping next to them, talking to them, stroking and fondling them, calling them by name, leading them on a leash to the fields, weeping when they fall sick or are injured, and feeding them with choice morsels from the family table.... The climax of pig love is the incorporation of the pig flesh into the flesh of the human host and of the pig as spirit into the spirit of the ancestors.... Pig love is honoring your dead father by clubbing a beloved sow to death on his grave site and roasting it in an oven dug on the spot. Pig love is stuffing fistfuls of cold salted belly fat into your brother-in-law's mouth to make him loyal and happy.

from Cows, Pigs, Wars and Witches

So, if you're a pig-lover, you need not feel guilty about eating pork sausages or bacon. The ultimate expression of your affinity may be to eat your pig.

Piggy People

"Then," said he, "the pigs are a race unjustly calumniated. Pig has, it seems, not been wanting to man, but man to pig."

from *The Life of Samuel Johnson*, James Boswell

Another expression of the affiliation between people and pigs is pig impersonation. Certain people, either through emulation or accident, are so unmistakably porcine in contour and character that they defy description in human terms — they are what we call "piggy people."

His beard was as red as any sow or fox,
And as broad as if it were a spade.
Right on the tip of his nose he had
A wart, and on it was a tuft of hair,
Red as the bristles of a sow's ears;
His nostrils were black and large.

description of the Miller
from the Prologue to
The Canterbury Tales,
Geoffrey Chaucer

Piggy people come in many varieties. There's the round, fat, pink person, with piggy eyes and a faint smile etched across his lips. He is the pantalooned dandy of the nineteenth century and the rotund rustic of the twentieth. Harmless, slow, and bemused, he plods through life. He's occasionally dozy, not very bright, but always beaming.

The gray and matted person who sports a mean look under drowsy eyelids is the lecherous hog, the drunken slob, the glutton, the shirker. (He doesn't resemble an actual pig; rather, he is like our worst misconception of the animal.) Behind that sneer lurks a pigheaded intelligence. Never underestimate a lazy hog.

The worldly wag is the Bluff King Hal of piggy people. Perky, scrubbed, trim of trotter, he buffoons and lampoons his way through life. He is the opportunist, the optimist, and the eternal winner; the pig of political cartoons and satire. No flies on him.

The incurably romantic boar is always crusading. Because he's afraid of nothing and nobody, he is a dangerous fighter. Lithe of muscle, eyes aglint, and mouth curling in a roguish grin, he is fiercely loyal to his friends. This tusker is a dark and deadly enemy. Sadly, old boars of the romantic school tend to become "crashers," spending most of their time reminiscing from the depths of leather chairs in gloomy club houses.

The epitome of security and love, and filled to the brim with the milk of human kindness, is the large, round, contented matriarch who sits happily at home, surrounded by swarms of squeaking children. She is mother earth and mother nature combined. Her soft, white skin glows, and her eyes, through half-closed lids, radiate porcine peace. Occasional grunts of pleasure indicate that tranquility is hers.

Swinish Satire

The artist notes that the cut along the dotted line refers to the slaughter of the pig, while the anchor stands for the hope that this might not happen. But the sausage and bacon indicate quite clearly the pig's inevitable future.

Given that so many of us think pink, it seems strange that pigs should be so ruthlessly personified, punned, and panned. The sad truth is, no other creature has been so extensively ridiculed. Although the irony and imagery is sometimes gentle and affectionate, by and large pigs are the scapegoats of our weaker traits. What is it that we are laughing at when we ridicule the pig? Could it be ourselves?

I'm sick and tired of hearing derhogatory remarks about pigs!

If your room is filthy, you say, "Pardon the mess in my room — I know it looks as though I'm living in a pig pen." Your room is in a mess because you are a mess. Please don't blame it on pigs!!

The only reason pigs in pigpens are filthy is because of management. Nothing that a good shovel won't fix.

You're sweating like a pig? Not true! Pigs do not sweat! They puff! They're trying to cool off.

Okay, if pigs aren't the filthiest beasts on earth, why do I see them out in meadows wallowing in mud — absolutely covered in the stuff? Mud protects their skin from the sun's rays. I'd like to see you standing naked in an open meadow.

When someone calls me a pig, I say thank you. Pigs have everything going for them: intelligence; clean living; good looks; and most importantly good taste. I can honestly say that any pig I've lent money to has paid me back. That is more than I can say about some people.

Thomas Hagey, publisher of
Playboar Magazine

Metaphorically Speaking

Pig abuse is all-pervasive. It is something we resort to almost unconsciously when we want to describe obese, selfish, grasping, uncouth layabouts. Indeed, the inclination to resort to a pig-metaphor can't be resisted....

Pearls Before Swine

Once upon a time, there was a young girl called Edweena, who lived with her wicked stepmother and stepsisters in a large house. The stepmother was called Madame Porcina, and the two stepsisters were named Chipolata and Nutella. Madame Porcina and her daughters were very ugly, with pink fat arms, pink fat legs, and snubby noses, and on Madame Porcina's chin sat a large brown mole from which sprouted two red bristles. They were all jealous of the slim, pretty, sweet-natured Edweena, and they made her life a misery by putting her to work in the kitchen and scullery. Not only did she have to cook and clean, but she also had to attend to the whims and wishes of Madame Porcina and the two sisters, who bossed her around from dawn till dusk. Edweena was forced to wear old tattered clothes, yet even dressed in rags she managed to look prettier than her stepsisters, and this made them all the crosser and nastier.

One of Edweena's duties was to get up at five in the morning and prepare the swill to feed the pigs who lived in the orchard. They were always pleased to see Edweena.

She would sneak some treats into the potato parings, and their gruntings and gentle snufflings were a great comfort to her. Harried as she was, she found the time to pat Mother Sow and scratch Old Tusker, and she helped the aged boar with his swill, for his tusks had been terribly neglected and got in his way.

One day, an invitation arrived at the house. The King had decided it was time for his son to find a wife, and was holding a large ball at the palace, to which all and sundry were invited, especially all the young ladies of the land. Madame Porcina fairly bristled with excitement. Here was a chance for her darlings to find a handsome husband. Chipolata and Nutella talked of nothing else for days. They also fought over the young Prince.

"If you don't stop guzzling, you greedy pig," said Nutella to Chipolata, "the Prince won't even look at you."

"Look who's talking, you lazy swine," retorted Chipolata. "Who wants a wife who lies hogging in bed all morning?"

And so they went on. They were very unpleasant young ladies.

Edweena also dreamed about the ball, and she plucked up the courage to ask Madame Porcina if she might go.

"You, go to the ball!" said Madame, with a petulant sniff of her nose, her bristles twitching with indignation. "Why, look at you, covered in hogwash, only fit for the pigsty. Get you gone, you dirty baggage."

The evening of the ball came round. Pandemonium ensued. Chipolata squeezed and puffed herself into a flouncy, pink-sequined gown and could hardly breathe. She had overeaten again. Nutella had her hair in curling tongs, and she screamed at Edweena to come and mend her buttons and the massive holes in her white stockings.

"Hurry, Edweena, you lazy bones," yelled Chipolata.

"My ponytail, you ninnyhammer," screeched Nutella.

"My pigtail, you dolt," bellowed Chipolata.

"My pompadour," raged Madame Porcina, wrestling with her powdered wig while she smudged her rouge and got lipstick all over her yellow teeth. Madame was by now a bundle of nerves and had already taken several swigs from the gin bottle, which she kept tucked under her bed. She looked flushed and dangerous as she swept her girls off into the carriage.

Edweena sat alone in the dark cold kitchen and burst into tears. Suddenly, the stillness of the room was shattered by an explosion and a flash of light. When the smoke cleared, Edweena saw a little old lady dressed in blue, with snow-white hair and twinkly beady eyes.

"Now, now, Edweena," she said. "Stop crying. I am your Fairy Godmother, and if you blow your nose and pull yourself together, you will go to the ball tonight."

"How can I go to the ball dressed in rags and smelling of pigs?" cried Edweena.

"Nothing wrong with pigs," said the little old lady. "It's just that hogwash you give them. Now, close your eyes and clap your hands, and pigs might fly."

Edweena closed her eyes and clapped her hands. When she opened them again, all the pigs from the orchard were inside the kitchen, running around getting things ready for her. There was tremendous excitement, and several of them squealed for joy. This was fun.

"You are going to live high off the hog tonight, Edweena," said one old boar, with a pleased grunt.

"You're going to be the belle of the ball," squeaked a young porker, who was smoothing out the ball gown with his trotter.

"In fact, you're going to go the whole hog, my dear," said a matronly old sow, who was in charge of the velvet cape and the jewelry.

"Take good care of these pearls. They've just been strung," said an excited young gilt. "And don't cast them before those swine," she giggled, pointing upstairs to where the stepsisters had their rooms.

All the pigs started chortling and stamping their trotters with glee. None of them liked Madame Porcina or her daughters.

"The old Madame was as drunk as a sow when she left tonight," muttered one old matron. "I don't know what the world is coming to. Drunk and driving in a carriage."

"That's enough, pigs," said the Fairy Godmother. "Remember the saying: 'too much gossip will bring one's pig to market' — and we want tonight to be a success for our Edweena, don't we?"

"Indeed we do, yes, yes," chorused the pigs, and efforts were at once renewed to dressing Edweena in all her finery. At last she was ready. Her gorgeous white ball gown was trimmed with lace, and a gleaming pearl necklace adorned her white throat. She was led by the pigs to a beautiful carriage shaped like an acorn, pulled by six handsome young piglets who were champing at their harnesses.

"Oh, thank you, Fairy Godmother," said Edweena. "Is this a dream, or am I really going to the ball?"

"You are going to the ball, my dear, and you are going to have the time of your life," twinkled the little old lady. "But you must remember one thing: at midnight, this magic spell will begin to wear off. Your beautiful clothes will change back into your rags, so you must leave and run home on the dot of midnight, whatever you do."

"Oh, I will, and thank you, Godmother, for everything. Thank you, dear pigs."

"Yes, yes," snorted the pigs. "Yes, yes, have a lovely time, Edweena."

"Save us a truffle or two," grunted Old Tusker, who remembered better days of sporting times and champagne and truffle parties.

"I'll try and make the swill extra special tomorrow, you kind pigs," said Edweena, planting a kiss on each of their snouts and giving Old Tusker a special scratch on his back, which made him grunt and snort with pleasure. Then she climbed into the acorn carriage.

"Trot along," said the Fairy Godmother. And away they went.

The piglets wasted no time galloping off into the night, and the next thing Edweena knew, she was walking up the palace steps, as if in a dream. The young Prince took one look at her and fell in love. He danced with no one else for the rest of the night. He had had a very boring evening up until she arrived. His toes had been trampled on by Chipolata, his epaulet torn by Nutella, and his head was spinning from the gin fumes exuded by Madame Porcina. Now he was obliviously happy. Not so Madame Porcina and the girls. They gnashed their teeth in impotent rage. They had not recognized Edweena, but this gorgeous young girl dancing with the Prince made them green with jealousy, and they muttered together in fury.

"What a guttersnipe, a road hog, a floozy," sneered Chipolata, darting scathing looks at Edweena.

"She's just a coarse, pimply, scrawny old sow," said Nutella, who suffered from pimples and automatically hated anyone who didn't. Madame Porcina called for a carriage and gathered up her herd of glowering girls.

"I will not stay to be insulted by an upstart young man who's got the wrong sow by the ear," she announced pompously and seized a final glass of champagne, which she guzzled before swaying down the palace steps.

The Prince was just asking Edweena her name, when the clock started to strike midnight. One. Two. The ominous sounds reverberated through the night.

"Oh, please excuse me, sir, but I must go," said Edweena, seizing her long white dress and running for the doors, where she tripped and fell. The white pearl necklace became unclasped and rolled down the steps. Edweena hobbled through the streets, feeling the dress slowly disintegrate into the familiar rags as the clock struck on. Cold and dirty, but happy, she crawled into her little bunk by the kitchen sink. When she fell asleep, she dreamed of waltzes, music, the glitter of the ballroom, the scrumptious dinner, and most of all the handsome Prince.

The Prince, meanwhile, was in a terrible stew. He tore around the castle like a hog in armor, searching high and low for the nameless beauty who had captured his heart. Racing down the palace steps, he came across the pearl necklace and recognized it as the one worn by his beloved. On picking it up, he realized what a small neck the wearer had. He vowed to himself there and then that he would search the countryside until he found his love, and that she would be the girl he would marry. He went and informed his father,

the King, of his plans. The King was a pompous old bore who had decided at the ball that the Prince should marry the Dowager Lady Troughenough's daughter, the young Lady Sowester, who had pots of money and a title to boot. But the Prince was pigheaded. After a mild tantrum, where he jumped up and down and called the Dowager a lump of pig iron, he stormed out, leaving his father sweating and puffing and puce in the face. The Prince then ordered his men to go to every household in the kingdom and try the pearl necklace on every neck of every woman and girl in the land, until his beloved was found.

And so it was that one fine morning, the Prince's retinue knocked on Madame Porcina's door.

"Madame," said the messenger. "Orders from his young Majesty, the Prince. All ladies are to try on this pearl necklace, and whoever has the neck that fits the necklace marries the Prince."

Chipolata and Nutella were hanging over the banister, eavesdropping as usual, and they nearly fell over into the hall in a flurry and scurry to try on the necklace. Madame Porcina, leering a coquettish grin, seized the necklace and, with shaking fingers, placed it on her sagging neck. The pearls disappeared into the folds of flesh.

"I'm sorry, Madame," spluttered the embarrassed messenger. "Next one please."

Chipolata was panting and out of breath. She had just made a pig of herself over a large box of chocolates and was fatter than ever. She stretched the necklace as tightly as possible round her throat, until she went purple and nearly choked. Nutella tried stretching her neck and clenching her jaw, but her Adam's apple protruded so much that the necklace wouldn't lie flat. The messenger bowed, placed the necklace back in its box, and prepared to depart. Then he noticed Edweena crouched down by the hearth.

"Ah, I see we have one more lady here," he said.

"That," choked Madame Porcina "is no lady. That is the servant, the scullery maid, the pig-keeper."

"Sorry," said the messenger, "Prince's orders. She must try on the necklace."

"Piffle and twaddle," snorted Chipolata.

"Bitter almonds, sour grapes, and raw onions," muttered Nutella.

Edweena crawled out from the hearth, covered with soot. The pig-swill remains were still in her lap. She bowed her head in shame and approached the messenger. He took the necklace from the box and placed it round her neck rather cautiously, as if he were afraid to get his hands grubby. But lo and behold! The necklace was a perfect fit. Edweena became transformed before their eyes. The rags and dirt disintegrated, and within seconds, a pretty, clean girl in a spanking white dress stood in front of them. The messenger jumped back and, gathering his wits together, plunged into a bow. "You are the Prince's beloved. Your wish is my command," he stammered.

The rest of the courtiers flopped to their knees, and the smallest page was sent scampering off to the palace to relate the good news to the Prince. Madame Porcina had a fit of convulsions and had to be revived with smelling salts. Nutella turned white and speechless, her eyes on stalks, and Chipolata blubbered into a pink frilly handkerchief. Suddenly the Fairy Godmother appeared.

"My dear," she said to Edweena, "no longer will you have to play pig-in-the-middle. You will make a lovely princess. Be happy with your Prince. And as for you, you battle-axes," she said, turning to Madame Porcina and her brood, "it's high time you dirtied your hands a bit." And she gave a little snort.

Immediately from outside the house, the sounds of squealing, snortings and grunting could be heard. The clamor got louder and louder and nearer and nearer. Then there came a tremendous crashing of tusks and clattering of trotters, and into the hall surged the pigs from the orchard, led by Old Tusker himself.

They charged straight for Madame Porcina and Nutella and Chipolata, rounded them up, and shuffled them out into the garden, where they chased them round and round the orchard. Then they pushed them headfirst into the troughs. And there the three harridans lay, entrenched in potato peels and banana skins.

Meanwhile, the Prince was jogging round the palace ramparts. When he heard the news, he was tickled pink, and jumped up and down for joy. Still in his track suit, he dashed off to see his beloved Edweena. Sure enough, she was the same girl he had danced with at the ball, and she looked as gorgeous as she had done that evening. Within a week, they were married, and they lived as happily ever after as pigs in clover.

Sarah Krzeczunowicz

Greedy Pigs

One cannot deny that pigs love to eat. And why not? Food is there to be enjoyed. The term "greedy pig" applies to those gluttons who will eat anything and everything in a manner that would make a pig blush. Take the Romans or the Edwardians, for instance. Twelve-course dinners were standard practice for such gorgers, who, when stuffed to the gills, would regurgitate their food by gently tickling the backs of their throats with feathers, so that they might start all over again. During the European Renaissance, aristocratic gourmands suffered from such serious bouts of indigestion due to overindulgence that they couldn't lie flat, and their four-poster beds overflowed with bolsters to support their bloated bodies.

Pigs turn up their snouts at such behavior.

Political Pigs

The political arena is the favorite battleground of the satirists. These sadistic gladiators have constantly favored the pig as the personification of greed, lust, stupidity, and stinginess. In political cartoons, the hog rarely plays center stage, but it has many interesting secondary roles.

THE MINISTERIAL SHANTY; OR, THE CAUCHON AT HOME.

Poetic Pigs

Poets especially enjoy using pigs when poking
fun at their contemporaries.
    The erotic affairs that you fiddle aloud
    Are as vulgar as coins of the mint.
    And you merely distinguish yourself from the crowd
    By the fact that you've put them in print.
    For your dull little vices we don't give a fig,
    It is this that we deeply deplore:
    You were cast as a common or garden pig,
    But you play the invincible bore.

To a Boy-poet of the Decadence, Owen Seaman

Alexander Pope, the father of the rhyming couplet, never minced his words, particularly when panning other wits:

Now wits gain praise by copying other wits
As one Hog lives on what another shits.

from *The Dunciad*

Pigassos

When artists seize the brush to daub their canvases with luscious ladies, are they reliving childhood fantasies? Often a subconscious attraction for the pig emerges in the brushwork, and the pig's form is then seen as highly desirable.

Consider the contours of a Botticelli, Rubens or Renoir portrait — the curves, the dimples, the pink and white buttocks, the ivory creases. Round, benign, and irresistibly rosy. Infinitely porcine.

Could it be that the current pig fever is really a latent nostalgia for the flesh? Where are the ample bosoms and fleshy thighs? Only to be found in pastures green? Is today's lean and hungry feline look destined for extinction? Are the fat folds due for a comeback? There are some who still hope. . . .

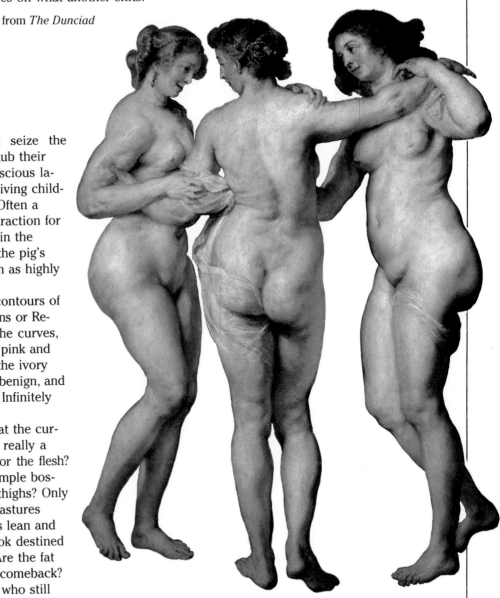

Male Chauvinist Pigs

Pigs are not as vainglo-rious as the archetypi-cal male chauvinist, but the comparison has stuck. The confident male chauvinist appears to enjoy the "pig" tag enormously. He is easily spotted at parties, sporting his piggy emblem embla-zoned on a dark tie. He is trying hard not to become an endangered species. Ac-tually, time and the mores are on his side.

"A male chauvinist pig is nothing more
    than a bore,
But he keeps me in business at least, the beast,"
    said the whore.

Sarah Krzeczunowicz

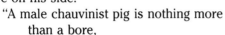

Drunken Swine

This metaphor is not only
offensive, it is downright
idiotic as well. Who on
earth would give a pig a
drink? A lonely lush desper-
ate for company?

The Pig Got Up and Slowly Walked Away

One evening in October when I was far from sober,
To keep my feet from wandering I tried.
My poor legs were all aflutter so I lay down in the gutter,
And a pig came up and lay down by my side.
We sang, "Never mind the weather just as long as we're together,"
Till a lady passing by was heard to say,
"All his self-respect he loses when such company he chooses,"
And the pig got up and slowly walked away.
Yes, the pig got up and slowly walked away,
Slowly walked away, slowly walked away.
Yes, the pig got up and then smiled and winked at me,
As he slowly walked away.

Benj. H. Burt (English version by Terry Sullivan)

Dirty Pigs

Oh, you unwashed
Little pig!
In what dirt
Did you dig?

The filth on you
Is inches deep!
You look. . .
A chimney sweep!

from *The No-Water Boy*,
Kornei Chukovsky

The pig is extremely clean, given the right surroundings. It is us who provide the dirt. Not satisfied with messing up our own homes, we must litter the pigsty, making it uninhabitable for the pig, who needs water to keep clean, and cool, fresh, juicy soil to rootle in.

It is hard to identify the courageous wild boar with the dull and sensuous hog of the following depiction. Pigs were meant to use their resources in the freedom of forests and, like anyone forced to live in filth and mire, will sink to depths of stupidity and sloth when deprived of a healthy environment.

The Hog, of all animals, is the most filthy. Rather fond of carrion than of flesh, he feeds on the dead carcasses of other animals, the refuse of the garden, barn, and kitchen. With this strange and indiscriminating taste, he has a ravenous appetite, and is fond of wallowing in the mire. His snout, too, which is well-adapted for ploughing up the wild carrot and other roots, renders him troublesome to the gardener and husbandman. The eyes of the Hog are remarkably small, and he has a dull and drowsy look. It is commonly remarked of him that he is useless during life and, like the miser, benefits only by his death.

from *The Juvenile Cabinet of Natural History* (1820)

Rude Pigs

We get the name of APER the Wild Boar from its savagery (*a feritate*) by leaving out the letter F and putting P instead. In the same way, among the Greeks, it is called SUA-GROS, the boorish or country pig. For everything which is wild and rude, we loosely call "boorish."

from *The Book of Beasts*,
a translation by T. H. White
of a twelfth-century
Latin bestiary

Whenever we attribute vulgarity to pigs, we expose our own lack of sensitivity. Queen Margaret proves herself more "boorish" than the target of her anger, Richard III, in Shakespeare's play. (Richard was not put off his hump-backed stride by Margaret's curse — perhaps he liked pigs.)

No sleep close up that deadly eye of thine,
Unless it be while some tormenting dream
Affrights thee with a hall of ugly devils!
Thou elvish-mark'd, abortive, rooting hog.

"Big Boar is Watching You"

Satirical humor reaches a sinister level in George Orwell's *Animal Farm*. The subtlety and stark simplicity of his "fairy story" make it a prophetic allegory.

Animal Farm is a satire of dictatorships, in which the pigs attain villainous heights. Old Major, a stout, regal-looking boar, helps the well-intentioned animals on the farm to organize a revolution to oust their cruel owner, Mr. Jones. When Mr. Jones is driven out, the vacuum of leadership is filled by the most intelligent beasts of the barnyard, the pigs.

Three pigs are quick to assume absolute authority: Snowball, the extrovert; Squealer, the persuader; and Napoleon, the silent manipulator. This pig hierarchy soon sets up rules for the

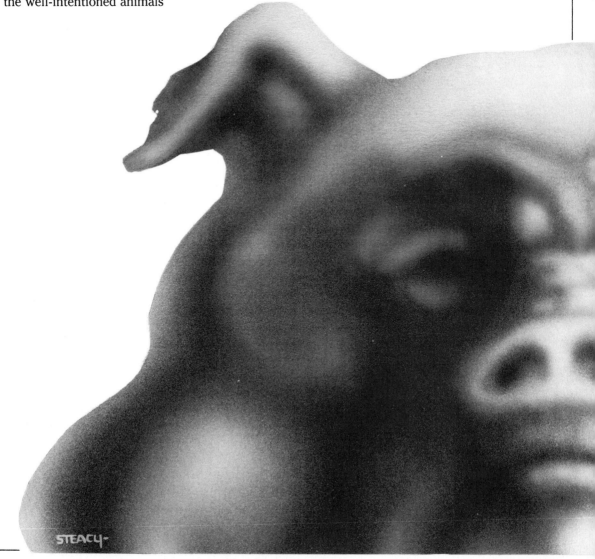

STEACY-

good of the community. The "Seven Commandments" forbid friendship with humans and the imitation of human vices. Most important, they ensure that all animals have equal rights.

Unfortunately, intelligence and good character don't always go together. Napoleon divides and conquers and quickly climbs to the top of the "party" heap, where he abuses his position of power. A reign of terror slowly starts — ex-cuses are made for promises not kept; animals who question the leadership disappear. The system rapidly deteriorates. The pigs not only imitate Mr. Jones, but their behavior is worse than his ever was. Squealer is seen tottering on his hind legs in a human caricature, and then Napoleon appears, tall and upright, brandishing a whip. The Seven Commandments are replaced by a single slogan, which states that some animals are now more equal than others.

The revolution turns full circle. The pigs become what they originally set out to supplant, and it is hard to distinguish which is man and which is pig.

Piglit:
Nursery to Sty

Piglitters

Pig ridicule belongs to the world of grownups. Children see nothing to mock or fear in pigs — just the opposite, in fact; they feel a sympathy that may have something to do with a remarkable resemblance.

A newborn — a little, pink cherub. All round and rosy and freshly bathed. A scrubbed, sweet-smelling picture of happiness. Sucking, gurgling, and occasionally scrambling. What a cuddly creature! Everyone wants to pick it up and talk nonsense to it; to squeeze it, tickle it, and hear it grunt.

As soon as she had made out the proper way of nursing it (which was to twist it up into a knot, and then keep tight hold of its right ear and left foot, so as to prevent its undoing itself), she carried it out into the open air. "If I don't take this child away with me," thought Alice, "they're

sure to kill it in a day or two: wouldn't it be murder to leave it behind?" She said the last words out loud, and the little thing grunted in reply (it had left off sneezing by this time). "Don't grunt," said Alice; "that's not at all a proper way of expressing yourself."

The baby grunted again, and Alice looked very anxiously into its face to see what was the matter with it. There could be no doubt that it had a *very* turn-up nose, much more like a snout than a real nose; also its eyes were getting extremely small for a baby: altogether Alice did not like the look of the thing at all. "But perhaps it was only sobbing," she thought, and looked into its eyes again, to see if there were any tears.

No, there were no tears. "If you're going to turn into a pig, my dear," said Alice, seriously, "I'll have nothing more to do with you. Mind now!" The poor little thing sobbed again (or grunted, it was impossible to say which), and they went on for some while in silence.

Alice was just beginning to think to herself, "Now, what am I to do with this creature when I get it home?" when it grunted again, so violently, that she looked down into its face in some alarm. This time there could be *no* mistake about it: it was neither more nor less than a pig, and she felt that it would be quite absurd for her to carry it any further.

So she set the little creature down, and felt quite relieved to see it trot quietly away into the wood. "If it had grown up," she said to herself, "it would have made a dreadfully ugly child: but it makes rather a handsome pig, I think." And she began thinking over other children she knew, who might do very well as pigs. . . .

from *Alice's Adventures in Wonderland*, Lewis Carroll

Unlike Alice's companion, the pigs most children first befriend are of the stuffed and cuddly variety. Pig love begins in the nursery cot, when mothers — among others — refer to the wriggly parts of babies' anatomy in a purely piggy way:

Little Pig,
Pillimore,
Grimithistle,
Pennywhistle,
Great big Thumbo, father of them all.

And

This little Piggy went to market,
This little Piggy stayed at home,
This little Piggy had roast beef,
This little Piggy had none,
And this little Piggy cried, "Wee, wee, wee,
I can't find my way home."

Youngsters are warned against getting their "piggies" cold if they don't wear slippers. And parents give "piggy backs," so that the world can be seen from higher up.

So pigs are not only on toddlers' minds, but piggy talk becomes part of their daily chatter. They learn to play "piggy-in-the-middle," a game where older children encircle a playmate with their hands joined, and while they twirl around, the trapped child dives — or tries to anyway — between their legs.

At bedtime, stories are read aloud about little pigs who dress in human clothes and who try very hard to keep out of mischief. As always, the new babysitter is given a hard time:

"What is her name?" asked William Pig.

"I don't like babysitters," said Benjamin.

"Oh," said Mrs. Pig, looking vague, "well, she's coming from the agency, so I'm not sure what her name is, but you're sure to like her."

"We didn't like the last one from the agency," grumbled Garth.

from *Mr. and Mrs. Pig's Evening Out*, Mary Rayner

There's sympathy, too, with that fed-up feeling from always being expected to obey:

"The darning-wool, the soap, the blue bag, the yeast — what was the other thing?" said Aunt Porcas.

"Wee, wee, wee!" answered Robinson.

"The blue bag, the soap, the yeast, the darning-wool, the cabbage seed — that's five, and there ought to be six. It was two more than four because it was two too many to tie knots in the corners of his hankie, to remember by. Six to buy, it should be — "

"I have it!" said Aunt Porcas. "It was tea — tea, blue bag, soap, darning-wool, yeast, cabbage seed. You will buy most of them at Mr. Mumby's. Explain about the carrier, Robinson; tell him we will bring the washing and some more vegetables next week."

"Wee, wee, wee!" answered Robinson, setting off with the big basket.

Aunt Dorcas and Aunt Porcas stood in the porch. They watched him safely out of sight, down the field, and through the first of the many stiles. When they went back to their household tasks they were grunty and snappy with each other, because they were uneasy about Robinson.

from *The Tale of Little Pig Robinson*, Beatrix Potter

And that free feeling when
the clothes come off:
They pulled off their clothes and ran on faster still,
Didn't even look back as they tore up the hill.
Breathless they ran through the old orchard gate—
For a roll in the mud the two pigs couldn't wait.
To be careless and free and to romp and to play
Was all that they wanted to do every day.

from *Pig Tale*, Helen Oxenbury

There are always exceptions
to wanting to have nothing
on, especially when it
comes to mud:
The small pig likes to eat,
and he likes to run
around the farmyard,
and he likes to sleep.

But most of all
the small pig likes to sit down

and sink down

in good, soft mud.

from *Small Pig*, Arnold Lobel

Children love *The Owl and the Pussy-Cat.* It rhymes (that being especially clever), and it feeds big imaginations with such wonderfully weird pictures of far-away lands and strange inhabitants:

> They sailed away, for a year and a day,
> To the land where the Bong tree grows
> And there in a wood, a Piggy-wig stood
> With a ring at the end of his nose....
> "Dear Pig are you willing
> To sell for one shilling
> Your ring?" Said the Piggy, "I will."

Edward Lear

Tales of daring pigs on secret missions also entice:

As I Looked Out

As I looked out on Saturday last,
A fat little pig went hurrying past.
Over his shoulders he wore a shawl,
Although it didn't seem cold at all.
I waved at him, but he didn't see,
For he never so much as looked at me.

Once again, when the moon was high,
I saw the little pig hurrying by;
Back he came at a terrible pace,
The moonlight shone on his little pink face,
And he smiled with a smile that was quite content.
But never I knew where that little pig went.

Author unknown

Naughty pigs, closer to home, are just as appealing:

Dan Pig

Dan Pig lived in a pigsty at the end of the garden. The pigsty had two rooms — an outside room and an inside room. The outside room had a trough in one corner for Dan Pig to eat his dinner out of, and a piece of wood in the other corner for Dan Pig to scratch his back against. The inside room had a big heap of clean straw in it for Dan Pig to sleep on. Near the sty grew an apple tree to make cool shade for Dan Pig in the summer.

One hot sunny day, Dan Pig stood in his outside room and waited for something to happen.

By and by, a dog came along the garden path. Dan Pig stood very still and quiet. The dog put his nose under the door of the sty.

"Hunc," said Dan Pig.

The dog was so startled he ran away into his kennel as fast as he could and stayed there for a long time.

That DID make Dan Pig laugh. "Hunc, hunc, hunc," he went. "Hunc, hunc, hunc!"

Then he stood still in

his outside room and waited for something else to happen.

By and by, a cat came along the garden path. Dan Pig stood very still and quiet. The cat put her nose under the door of the sty.

"Hunc!" said Dan Pig.

The cat was so startled she ran away as fast as she could, right up to the top of the apple tree, and stayed there for a long time.

That DID make Dan Pig laugh. "Hunc, hunc, hunc," he went. "Hunc, hunc, hunc!"

Then he stood very still in his outside room and waited for something else to happen.

By and by, Missis came along the path. She carried Dan Pig's dinner in a bucket. Dan Pig stood very still and quiet. Missis opened the door of the sty.

"Hunc!" said Dan Pig. Then he took the corner of Missis's apron into his mouth and pulled it.

"All right, Dan Pig," said Missis, "now leave my apron alone and have your dinner." And she tried to pull the apron out of Dan Pig's mouth. Dan Pig pulled hard, and Missis pulled hard. But Dan Pig pulled harder, and suddenly Missis's apron

came right off and fell on the floor of the sty.

That DID make Dan Pig laugh. "Hunc, hunc, hunc," he went. "Hunc, hunc, hunc!"

"There now, look what you've done," said Missis crossly. "Whatever has come over you today?"

Then Dan Pig saw that the door of the pigsty was open, so he ran right out!

"Come back at once, Dan Pig!" called Missis, and she began to run after him.

But Dan Pig knew she couldn't catch him, and he ran all round the garden three times before he went back into the sty. Missis ran in after him, all out of breath, and shut the door — bang!

"There," she said, "Now you just eat up your dinner, Dan Pig." And she filled up his trough with a nice dinner of potatoes and gravy.

"I don't want this dinner," said Dan Pig. "I don't want potatoes and gravy. I want nice juicy green grass." And he put his nose into the trough and began to push his dinner about. He pushed it out of the trough and all over the floor of the sty.

That DID make Dan Pig laugh! "Hunc, hunc, hunc," he went. "Hunc,

hunc, hunc!"

Then Dan Pig lay down and rolled over and over. When he stood up, there were pieces of potato all over his back and on his head and in his ears, and gravy was running down his nose.

"Dan Pig, Dan Pig, I'm surprised at you!" said Missis. "I don't know what's come over you today." And she picked up the bucket and went away and shut the door of her house — bang!

When she was gone, Dan Pig felt very uncomfortable. A piece of potato fell into his eye. Gravy was tickling his nose. He felt so hot.

By and by, he heard Missis coming back along the path. Dan Pig stood very still and quiet; Missis opened the door of the sty, but he didn't say a word! Missis was carrying a bucket of water and a big scrubbing brush like the one she used to scrub the kitchen floor.

"Now, you stand still, Dan Pig," said Missis. "I've never seen such a dirty pig in all my life."

She dipped her brush into the water and began to scrub Dan Pig's back. Dan Pig stood very still. Missis scrubbed away all the dinner from Dan Pig's back and head and

nose. And then she lifted up his ears and scrubbed around those. Dan Pig liked that.

When he was clean again, Missis threw a bucket of water all over him and said: "Now shake yourself." And Dan Pig did. Then she got more water and a long broom, and she washed and brushed the floor of the sty till it was clean again. Then she went away, and Dan Pig stood drying in the sun.

By and by, Missis came back. Her apron was full of nice juicy green grass.

She put it into Dan Pig's trough, and he ate up every bit. He had just finished, when a little green apple fell off the apple tree. It fell right into Dan Pig's trough!

That DID make Dan Pig laugh! "Hunc, hunc, hunc," he went. "Hunc, hunc, hunc!" And he ate that little sour green apple up too, stalk and skin and pips and all.

Then he went into his inside room. He lay down on his heap of straw and fell fast asleep, snoring.

Peggy Worvill

The delights of destruction go hand in hand with the satisfaction of construction. Forts are rapidly built up and knocked down during daily battle with the foe (the boy or girl next door) and the cleverest strategist comes out the winner — like the third little pig:

The Story of the Three Little Pigs

Once upon a time there was an old Sow with three little Pigs, and as she had not enough to keep them, she sent them out to seek their fortune.

The first that went off met a Man with a bundle of straw, and said to him, "Please, Man, give me that straw to build me a house," which the Man did, and the little Pig built a house with it. Presently came along a Wolf and knocked at the door and said, "Little Pig, little Pig, let me come in."

To which the Pig answered, "No, no, by the hair of my chinny chin chin."

"Then I'll huff, and I'll puff, and I'll blow your house in!" said the Wolf. So he huffed, and he puffed, and he blew his house in and ate up the little Pig.

The second Pig met a Man with a bundle of furze, and said, "Please, Man, give me that furze to build a house," which the Man did, and the Pig built his house. Then along came the Wolf and said, "Little Pig, little Pig, let me come in."

"No, no, by the hair of my chinny chin chin."

"Then I'll puff, and I'll huff, and I'll blow your house in!" So he huffed, and he puffed, and he puffed, and he huffed, and at last he blew the house down and ate up the second little Pig.

The third little Pig met a Man with a load of bricks, and said, "Please, Man, give me those bricks to build a house with." So the Man gave him the bricks, and he built his house with them. So the Wolf came, as he did to the other little Pigs, and said, "Little Pig, little Pig, let me come in."

"No, no, by the hair of my chinny chin chin."

"Then I'll huff, and I'll puff, and I'll blow your house in."

Well, he huffed, and he puffed, and he huffed, and he puffed, and he puffed, and he huffed; but he could *not* get the house down. When he found that he could not, with all his huffing and puffing, blow the house down, he said, "Little Pig, I know where there is a nice field of turnips."

"Where?" said the little Pig.

"Oh, in Mr. Smith's home-field; and if you will be ready tomorrow morning, I will call for you, and we will go together and get some for dinner."

"Very well," said the little Pig, "I will be ready. What time do you mean to go?"

"Oh, at six o'clock."

Well, the little Pig got up at five and got the turnips and was home again before six. When the Wolf came, he said, "Little Pig, are you ready?"

"Ready!" said the little Pig, "I have been and come back again and got a nice pot-full for dinner."

The Wolf felt very angry at this, but thought that he would be *up* to the little Pig somehow or other; so he said, "Little Pig, I know where there is a nice apple tree."

"Where?" said the Pig.

"Down at Merry-garden," replied the Wolf; "and if you will not deceive me, I will come for you at five o'clock tomorrow, and we will go together and get some apples."

Well, the little Pig woke at four the next morning and bustled up and went

off for the apples, hoping to get back before the Wolf came; but he had farther to go and had to climb the tree, so that just as he was coming down from it, he saw the Wolf coming, which, as you may suppose, frightened him very much. When the Wolf came up, he said, "Little Pig, what!

Are you here before me? Are they nice apples?"

"Yes, very," said the little Pig. "I will throw you down one." And he threw it so far that, while the Wolf was gone to pick it up, the little Pig jumped down and ran home.

The next day, the Wolf came again and said to the little Pig, "Little Pig,

there is a fair in the town this afternoon. Will you go?"

"Oh, yes," said the Pig, "I will go. What time shall you be ready?"

"At three," said the Wolf.

So the little Pig went off before the time, as usual, and got to the fair and bought a butter churn and was on his way home with it, when he saw the Wolf coming. Then he could not tell what to do. So he got into the churn to hide, and in doing so turned it round, and it

began to roll, and rolled down the hill with the Pig inside it, which frightened the Wolf so much that he ran home without going to the fair.

He went to the little Pig's house and told him how frightened he had been by a great round thing that came down the hill past him.

Then the little Pig said, "Hah! I frightened you, did I? I had been to the fair and bought a butter churn, and when I saw you, I got into it and rolled down the hill."

Then the Wolf was very angry indeed and declared he *would* eat up the little Pig and that he would get down the chimney after him.

When the little Pig saw what he was about, he hung on the pot full of water and made up a blazing fire and, just as the Wolf was coming down, took off the cover of the pot, and in fell the Wolf. And the little Pig put on the cover again in an instant, boiled him up, ate him for supper, and lived happy ever after.

Traditional

The Market Merry-Go-Round

To market, to market, to buy a fat pig,
 Home again, home again, dancing a jig;
Ride to market to buy a fat hog,
 Home again, home again, jiggety-jog.

The next stage in our piggy education is learning that pigs either go to or come from markets. Fortunately, the actual meaning of "going to market" is not divulged — it's simply a happy, bustling place just around the corner.

Most pigs are ready and willing to make the short trip. Some trot there with their lunches packed in knapsacks.

Others like to go in a horse-drawn wagon.

Still others need to be led, and not all pigs are eager to come home:

Little Dame Crump

Neat little Dame Crump, with her little hair broom,
One morning was sweeping her nice little room,
When casting her little gray eyes on the ground,
In a sly little corner a florin she found.

The Dame picked it up and exclaimed with surprise,
"How lucky I am! Oh, dear me, what a prize!
To market I'll go, and a pig I will buy,
And little John Gubbins shall build it a sty."

She put on her hat, and she tucked up her gown
And locked up her house and set off to town;
Across the green meadows she took her glad way,
Where the hedges were white with the blossoms of May.

At the market arrived, she looked well about,
The very best pig she could find to pick out:
A white one she chose, and a bargain she made,
And only a florin for Piggy she paid.

Her money thus spent, she was puzzled to know
How both would get home if the pig wouldn't go;
And fearing the creature might play her a trick,
She bought there, to drive him, a little crab-stick.

Through streets and through highways the little Dame went,
With Piggy's behavior at first quite content.
Till, just as they came to the banks of a brook,
An obstinate fit little white Piggy took.

He stood still and grunted, and on wouldn't go—
Now, wasn't he naughty to tease the Dame so?
She, finding that coaxing and scolding were vain,
Used her stick on his back till he went on again.

But, oh, not for long! At the foot of the hill,
Where the pathway runs up the neighboring mill,
Once more he refused to proceed on his way
And even the stick would no longer obey.

"Now, what shall I do?" said the poor little Dame,
"Night draws on apace; it is really a shame!
Well, Piggy, since thus you're resolved to stand still,
I'll leave you and go up for help to the mill."

Piggy gave just a grunt; as if he would say,
"Pray do as you like, I shall have my own way."
So she went to the mill and borrowed a sack,
Which she popped Piggy in and placed on her back.

She carried him thus to his neat little sty
And made him a litter of straw, fresh and dry;
With a handful of peas the Piggy was fed
And soon fell asleep on his snug little bed.

The Dame, very tired, as you well may believe,
Was glad little Pig to his slumbers to leave.
She went straight to bed, and she put out the light,
(But first said her prayers)—here we'll wish her "Good night."

Traditional

Classical Pigs

The art of writing for children is to make the pig more than just a pink roly-poly who lies idle behind sty doors; it is to give the animal depth and spirit; it is to invent a personality so real and memorable that the reader is changed by coming in contact with it.

In *Charlotte's Web*, E. B. White created such a character — the incomparable Wilbur. What makes Wilbur a classic is his ordinariness; children recognize him immediately. He has nothing to complain about in his life; he has food in his trough, and a friend in Fern (the girl on Mr. Zuckerman's farm who bottle-fed him when he was tiny and visits him in the barn quite frequently). But Wilbur is still lonely. He goes round the barn looking for a friend. He tries the goose, the lamb, and Templeton, the rat, to no avail. Just at the point when he is reaching absolute despair, he hears a voice from somewhere in the rafters, offering to be his companion. It takes him a while to locate the source of the voice, for it belongs to a spider called Charlotte, whose web stretches across the upper part of the barn's doorway. As being friends with a spider is not exactly what Wilbur has in mind,

getting to know Charlotte and her different way of life is his next task. No longer depressed, he puts on some weight and generally enjoys his lot in the barn much more. But when one of the sheep warns him that he is being fattened for a purpose, he becomes hysterical.

Charlotte calms him down and tells him that he is not going to die — she is going to save him. This bewilders Wilbur, as it seems impossible for a spider to have any influence on his supposed fate, and he keeps asking Charlotte about her plans. She always answers that she is working on it, and that is as far as it goes.

It is on a foggy morning that Lurvy, the man who feeds Wilbur, notices an extraordinary thing. There, in Charlotte's web, are the words SOME PIG, and this causes quite a furor at the farm. Word spreads to adjoining farms and throughout the neighborhood that a very special pig is at the Zuckerman's Farm:

"You notice how solid he is around the shoulders, Lurvy?"

"Sure. Sure I do," said Lurvy. "I've always noticed that pig. He's quite a pig."

"He's long, and he's smooth," said Zuckerman.

"That's right," agreed Lurvy. "He's as smooth as they come. He's some pig."

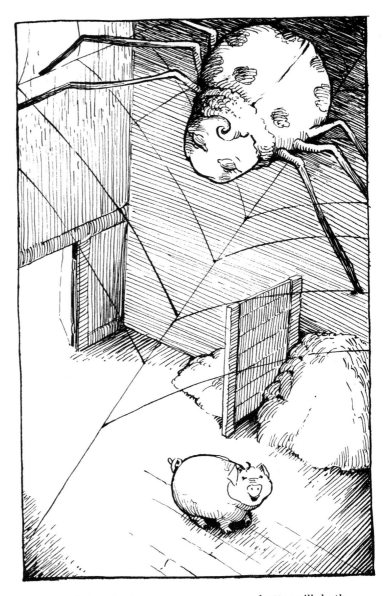

Charlotte works on her promise to save Wilbur's life. Toiling ceaselessly for long, long hours, she weaves the word TERRIFIC, and Wilbur grins beneath it.

And so Wilbur becomes famous:

The news of the wonderful pig spread clear up into the hills, and farmers come rattling down in buggies and buckboards, to stand hour after hour at Wilbur's pen admiring the miraculous animal.

All said they had never seen such a pig before in their lives.

He is to be shown at the County Fair as ZUCKERMAN'S FAMOUS PIG, and the morning of the Fair he is given the ultimate in luxury — a buttermilk bath: Wilbur stood still and closed his eyes. He could feel the buttermilk trickling down his sides. He opened his mouth and some buttermilk ran in. It was delicious. He felt radiant and happy. When

Mrs. Zuckerman got through and rubbed him dry, he was the cleanest, prettiest pig you ever saw. He was pure white, pink around the ears and snout, and smooth as silk.

Of course, all the animals from the barn have to be part of the great event, and they manage to get themselves smuggled along in the truck. That night, when all the animals are bedded down, Charlotte's new web complete, she assures Wilbur that he never again will have anything to worry about. Now he is SOME PIG, and ideas of slaughter will never be contemplated again. As Wilbur dozes off, he realizes that Charlotte is secretly at work on something else, but after inquiring, he is only told that what she is doing is for herself for a change, and that he will know about it in the morning.

The following morning, Wilbur awakens to witness Charlotte's "magnum opus," as she calls it. Her egg sac contains 514 little spider eggs. Wilbur feels proud of Charlotte, yet he guesses that there is something wrong. Charlotte tells him she is tired and "languishing," but that he shouldn't worry himself, as his big day as Zuckerman's Famous Pig at the Fair is still ahead of him. Charlotte has been

When Wilbur finally comes out of his crate for working overtime — she has once again woven a word for Wilbur in her web, except this time it reads HUMBLE. Wilbur loves it. public viewing, he blushes. He stands perfectly still and tries to look his best:

"This magnificent animal," continued the loud speaker, "is truly terrific. Look at him, ladies and gentlemen! Note the smoothness and whiteness of the coat, observe the spotless skin, the healthy pink glow of ears and snout."

"It's the buttermilk," whispered Mrs. Arable to Mrs. Zuckerman.

"Note the general radiance of this animal! Then remember the day when the word "radiant" appeared clearly on the web. Whence came this mysterious writing? Not from the spider, we can rest assured of that. Spiders are very clever at weaving their webs, but needless to say spiders cannot write."

"Oh, they can't, can't they?" murmured Charlotte to herself.

"Ladies and gentlemen," continued the loud speaker, "I must not take any more of your valuable time. On behalf of the governors of the Fair, I have the honor of award-

ing a special prize of twenty-five dollars to Mr. Zuckerman, together with a handsome bronze medal suitably engraved, in token of our appreciation of the part played by this pig — this radiant, this terrific, this humble pig — in attracting so many visitors to our great County Fair."

At this point Wilbur faints, obviously with the excitement of it all. There is a lot of hullabaloo, but he gets to his feet soon after, just in time to have his picture taken.

Wilbur and Charlotte have a talk that night. Wilbur has found not just a friend, but an irreplaceable friend, and he gushes his appreciation to Charlotte. He asks her why she has done so much for him:

"You have been my friend," replied Charlotte. "That in itself is a tremendous thing. I wove my webs for you because I liked you. After all, what's a life, anyway? We're born, we live a little while, we die. A spider's life can't help being something of a mess, with all this trapping and eating flies. By helping you, perhaps I was trying to lift up my life a trifle. Heaven knows, anyone's life can stand a little of that."

It isn't long after this that Charlotte informs Wilbur that she isn't going back to the farm with him. She has only two days to live, and her strength is fading fast. Overcome with grief, Wilbur sobs uncontrollably, but faces the facts by bribing Templeton to climb the wall and rescue Charlotte's egg sac, so that he can at least take that home, where it surely belongs. Wilbur and Templeton quietly leave a deserted fairground, and Charlotte, after waving feebly good-by with two legs, passes away.

Wilbur returns home with a bronze medal of honor around his neck and Charlotte's egg sac in his mouth. Winter passes, and when the last old strands of Charlotte's web have finally blown away, a tiny spider is born. It looks just like Charlotte. Wilbur is exceedingly excited. He starts chatting to the young spiders as they crawl from their sacs. But it isn't long — especially with the warm spring weather approaching — before the spiders take flight, and Wil-

bur is distressed at the thought of losing his new friends. Fortunately, however, three stay behind, making their webs, as their mother had done, in the barn doorway.

They are called Joy, Aranea, and Nellie, and Wilbur is filled with happiness. As the years pass, Charlotte's offspring come and go in the barn, but ...

> Wilbur never forgot Charlotte. Although he loved her children and grandchildren dearly, none of the new spiders ever quite took her place in his heart. She was in a class by herself. It is not often that someone comes along who is a true friend and a good writer. Charlotte was both.

The basic element of *Charlotte's Web* — is the yearning for a "best friend" — touches children of all ages. When we were young, we all wanted and hunted for a best friend, someone loyal who could be counted on to keep a secret. Invariably, best friendships endured little more than a week or two, but they were intense and binding while they lasted.

Pooh Bear's relationship with the impulsive imp, Piglet, reveals a more sophisticated form of real friendship. Piglet may not be Pooh's *best* friend, but he is usually reliable and open for an adventure. Piglet's appeal lies both in his timidity and his attempts to cover that timidity up.

Suddenly Winnie-the-Pooh stopped and pointed excitedly in front of him. "Look!" "What?" said Piglet, with a jump. And then, to show that he hadn't been frightened, he jumped up and down once or twice in an exercising sort of way.

From "Pooh and Piglet Go Hunting and Nearly Catch a Woozle", *Winnie-the-Pooh*, A. A. Milne

Romantic Runts

By the time adolescence strikes, it is a relief to know that others share similar heartaches:

The Story of the Pig and the Lion — A Fable

Once upon a time, there was a little pig. She lived in a green meadow and liked to romp in the lush grass. She loved the sun to shine but didn't mind too much when it didn't.

The other animals in the surrounding fields were quite friendly and helped her, if she needed help, and put their heads over the fence, if she needed company. Altogether, she was quite happy with her lot.

It was in the early autumn that things changed for her. One of the cows in the neighborhood told her of a visitor. The cow didn't say too much, only that his name was Simon, and he wanted company. Lucy the Pig asked the cow why she couldn't serve him just as well, and the cow said she was too occupied with a horse from the farm next door. When Lucy the Pig got back to her sty, she lay in the hay
and wondered why.

Next day, she was snorting and gobbling up some scraps, when she realized that there was shuffling on the other side of the trough. To her amazement and surprise, she noticed two big blue eyes staring at her, surrounded by a huge mane.

"Hello, I'm Simon the Lion."

"Oh!" said Lucy — she'd never seen a lion in her life before. Of course, she remembered talking about them when she was a young piglet, but now she couldn't believe her eyes!

"Do you want to go rolling in the hay?" said the big lion, grinning.

What cheek, thought Lucy the Pig, to suggest such a close, friendly thing when they had just met. "No thank you," she said sharply.

"Why not?" said Simon.

"Because it's not right." But Lucy was slowly becoming intrigued by this stranger, and she added, "Maybe another time."

"All right," Simon said, seemingly satisfied. "Let's go for a walk instead."

So off went the two of them across the meadow, chatting away. Lucy observed Simon carefully out of her beady black eye. He was a big lion and had lovely legs and paws, and he was gracious and elegant, definitely a king of the animal kingdom. She was fascinated but kept her cool, looking down at her little trotters.

The weeks went by. Simon came to see her almost every day, and they lay in the tall green grass of the meadow and talked. Simon told Lucy of his life, and she felt sorry for him when he looked extra sad as he told her about a lioness he had lost to another lion. She also felt muddled about why he was interested in a pink pig as opposed to his normal breed, but she thought it better to keep quiet. Besides, he was a very open, loving lion, and he would tell her in due course.

One day, after a huge supper of slop — the farmer was extra generous that day — they rolled in the hay together. It was such a delicious experience for them both that they rolled every day thereafter. And Lucy the Pig thought she must be in love with Simon the Lion.

It was beginning to get cold, and winter was on its way. Simon was starting to shiver and asked Lucy whether she'd come with him back to the jungle where he lived. But Lucy didn't know what to

say. Obviously Simon needed to be back in a warm country, but could *she* take the heat, seeing she wasn't used to the climate. She looked out of the sty door and across the fields. The wind howled noisily. Could she leave her old friends on the farm? It would be so difficult.

At this point, she thought she should consult the cow, so off she trotted through the mud to the milking shed.

"It really depends on how important the lion is to you," said the cow. "Why don't you wait a while and think about it?"

So Lucy, having got no further, returned to her sty

and wondered why.

Simon the Lion had become rather distant. Lucy tried to pick his brain and see if anything was wrong, but he seemed at a loss for an answer and said he should go back to the jungle alone. Lucy was upset about this, but she thought it was probably for the best, as her Pig Mother had told her so many times.

So on a cold wintry day, Simon slipped away. And Lucy sat in her sty
    and wondered why.

L. Hollingshead

Finding someone to love is rewarding for any pig or person; however, if a person falls for a pig and expects a traditionally human response, the problems involved will prove insurmountable. Pigs basically like their own kind and are not always co-operative.

Author unknown

THERE WAS A LADY LOVED A SWINE

1. There was a la-dy loved a swine, "Ho-ney!" said she;
2. "I'll build thee a sil-ver sty, Ho-ney!" said she;

"Pig - hog, wilt thou be mine?" "Hunc!" said he.
"And in it thou shalt lie!" "Hunc!" said he.

3. " Pinned with a silver pin,
Honey !" said she ;
"That thou mayest go out and in,"
" Hunc!" said he.

4. " Will thou have me now,
Honey?" said she ;
" Speak, or my heart will break,"
" Hunc!" said he.

It takes a while to warm up to parties. Shyness makes meeting others hard work and finding common ground frustrating:

The Warthog Song

The jungle was giving a party,
a post-hibernation ball —
the ballroom was crowded with waltzing gazelles,
gorillas and zebras and all.

But who is this animal,
almost in tears,
pretending to powder her nose:
A poor little warthog who sits by herself
in a pink satin dress with blue bows.
Again she is nobody's choice as she
sings in a sad little voice,

Wart-hog   I'm a Wart-hog___ un-der-neath"

"Take your partners for a Ladies' Excuse Me."
Excited and radiant she runs on the floor
to join the furor and fuss.
She taps on each shoulder and says, "Excuse me,"
and each couple replies, "Excuse us."
Then having no manners at all,
they sing as they dance round the hall.

Chorus:
And no one ever wants to court a warthog,
though a warthog does her best.
Her accessories are dazzling for a warthog,
she is perfumed and daringly dressed.
We know her "these and those" are like Marilyn Monroe's,
her gown is just a scintillating sheath,
but somehow fails to please,
'cos everybody sees that she's a warthog, just a warthog,
she's a warthog, underneath.

Head hanging, she wanders away from the floor,
this warthog whom nobody loves —
and stops, in amazement, for there at the door
stands a gentleman warthog impeccably dressed,
in the act of removing his gloves.
His fine chiseled face seems to frown
as he looks her first up and then down.

Chorus:
"I fancy, you must be, er, a sort of a warthog —
though for a warthog you look a mess.
That make-up's far too heavy for a warthog.
You could have chosen a more suitable dress.
Did you have to dye your hair?
If that's perfume, give me air.
I strongly disapprove of scarlet teeth —
but let us take the floor, 'cos I'm absolutely sure
that you're a warthog, just a warthog,
the sweetest little, neatest little,
dearest and completest little warthog
underneath!
(kiss kiss)   Michael Flanders and Donald Swann

Happily, things usually
work out for the best,
with a little persistence:

The Story of the Tell-Tale Tail

One morning Sir Warthog uneasy he felt.
He emerged from the bush and adjusted his belt.
He straightened his tail till it rose high and fair
(A family habit he was proud to share).
He polished his tusks and he groomed his gray hair
And wondered: "What's wrong on a morning so rare?
I've found lovely truffles, I've bathed in the pool,
Took a snooze in the shade and found it so cool.
I've sung day and night to the Sun and the Moon,
And discovered a mud hole for the heat after noon.
So why am I restless? There's nothing I lack.
I have cacti and baobabs for scratching my back,
And more food and drink than I ever could use.
So what is this need? My heart's one big bruise!"
He went to the Turtle, so agéd and wise
And snuffled and grunted and asked 'mid his sighs:
"What causes my bother? What's wrong with my life?"
"Dear friend," came the answer, "your need is a wife."
"Me? A wife! Oh, dear! That's so easy to say.
Talk about searching for a needle in hay!
The bush is so thick; it's so beastly long.
Where is the maid I can woo with my song?"
Yet bravely Sir Warthog went forth on his quest,
And he searched and he hunted — he knew no rest.
But as day followed day his morale went down.
(His sadness was noted by beasts all around.)
Till one day when weary and downcast he rested,
He saw something strange from where he was nested.
He could see a tail on a bush straight ahead,
Standing up in the air!! He scratched at his head.
"A bush with a tail. Something warty in there."
And he jumped in the bush as quick as a Hare.
Now the bush has two tails and the tale that they tell
Is told every day where'ere lovers dwell.
But for those, unlike us, who have not yet guessed,
The bush now with many young Warthogs is blessed.

K.J. Krzeczunowicz

On cool summer days
I gaze
At the pigs.

As they scuffle, and snuffle
I sit
With the pigs.

As they trot and grunt
I laugh
With the pigs.

What a time I'm having
Playing
With pigs.

But I've noticed
They're growing
These little pigs.

Even as grownups
They're pretty
As pigs.

No more cuddles and coos
Can I have
With these pigs.

But I'll love them
Forever
My friends, the pigs.

L. Hollingshead

Pigs Is Equal

There are pigs and there are pigs — they're not all captivating. Yet once in a lifetime a special pig comes along, a pig with whom to spend time, chat, share innermost secrets, relax, and enjoy a good laugh. No one need analyze the closeness of this com-panionship; it is simply there to be enjoyed.

The country is the place to put everything into per-spective. The leisurely pace, the morning sun, the squelching mud, the friendly sty, the rustle of hay from within, the bird on the roof-top, the scrape of the wooden door, and then, in-side, the sense of peace that feels like home. While sitting there, pondering the pig, a basic need for secur-ity is fulfilled. Apart from occasional grunts, there's nothing much to say. It's all understood.

To Love and To Cherish

The actual lines of a pig (I mean of a really fat pig) are among the loveliest and most luxuriant in nature. The pig has the same great curves, swift and yet heavy, which we see in rushing water or in rolling cloud.

G. K. Chesterton

The most angelic sight in the world is that of a child asleep. Tenderness and tranquility emanate from the sleeping face. Every muscle is relaxed; the hair slightly askew on the pillow, the mouth partially open, and the breathing soft and even. Occasionally, a flutter ripples across the softness as pleasant dreams float through the night. The most active demon by day is transformed by dusk into a peaceful cherub.

That vision is only matched by a pig asleep in its sty. Amid the clamor and clatter of the barnyard, the pig can settle in the straw and be asleep in seconds, oblivious to all distractions. Pigs need room. Sleep is not a scrunched-up affair in a corner of the sty — sleep is a stretched-out, delicious experience, which is well-planned and savored. The straw is snuffled and shuffled into place, well in the middle of the pen, so that the sleeper can stretch and languish at leisure, until sleep closes up the beady eyes. Then the trotters slowly relax, the ears give a final twitch and settle themselves for the night, the mouth droops, and the faintest gleam of a tusk can be seen smiling through the sleep. A bit of snout is tucked under the straw for added warmth, and the tail is nowhere to be seen. Couples are especially devoted. They cuddle up next to each other, trotters entwined and snouts gently resting alongside one another. Very little can disturb the bliss of bedtime.

To sleep, perchance to dream ... Dreams of cool water, fresh green grass, and orchards full of apples may cause a faint stirring, and certainly thoughts of truffles and mushrooms will start eyelids flickering and mouths drooling, but dreams come and go, and sleep is heavy.

Poetry in motion ... The moment a pig wakes up, peace and quiet cease. The pig sty becomes a hive of activity. A pig takes nothing for granted; everything must be inspected, shuffled, pushed, nosed, rootled, and thoroughly checked-out before the pig will stop and settle for a few minutes of back scratching against the side of the trough. Then the straw is tossed around a little more, and the ground snuffled for any tidbit that may have been overlooked the day before. After a drink and a quick clean-up of the already emptied trough, a lookout position is established, with trotters perched on the gate and snout peering hopefully over the top. Breakfast always seems late, but there's plenty of neighborhood chatter and gossip to while away the time — squeals from piglets dashing between mothers' legs, grunts from still-sleepy sows, and determined snorts from lonely boars. Breakfast arrives. Excitement and curiosity reach fever pitch. A new person is bringing the swill. There is more sniffing and a lot of investigating. Does the slop-bearer pass the test? Then silence, except for sounds of swooshing and slurping as the troughs are licked clean several times, overturned, checked, and rechecked. A little nap to digest breakfast is followed by either a mud wallow or cold shower to cool off. The rest of the day is spent in more scratching, rootling, and tidying up, until sleep comes round again at night, and all is quiet.

Pigs grunt in a wet wallow-bath, and smile as they snort and dream. They dream of the acorned swill of the world, the rooting for pig-fruit, the bag-pipe dugs of the mother sow, the squeal and snuffle of yesses of the women pigs in rut. They mud-bask and snout in the pig-loving sun; their tails curl; they rollick and slobber and snore to deep, smug, after-swill sleep.

from *Under Milk Wood*, Dylan Thomas

To Comfort, Honor and Keep

You cannot separate a devoted pig-man from his pig. So deep and intense is the relationship that a true pig-lover can instinctively smell out another pig-man.

Famous vet James Herriot of *All Creatures Great and Small* is obviously paid the highest compliment when he visits Mr. Worley, the local pubkeeper, whose sow has fallen ill:

"I can see you like pigs," said Mr. Worley as I edged my way into the pen.

"You can?"

"Oh yes, I can always tell. As soon as you went in there nice and quiet and scratched Queenie's back and spoke to her, I said, 'There's a young man as likes pigs'."

After Mr. Herriot has taken care of Queenie, Mr. Worley goes on.

"Thank ye, thank ye, I'm very grateful." He shook my hand vigorously as though I had saved the animal's life. "I'm very glad to meet you for the first time, Mr. Herriot. I've known Mr. Farnon for a year or two, of course, and I think a bit about him. Loves pigs does that man, loves them. And his young brother's been here once or twice — I reckon he's fond of pigs, too."

"Devoted to them, Mr. Worley."

"Ah yes, I thought so. I can always tell."

Mr. Worley has a whole harem of pigs: They were all Tamworths, and whichever door you opened you found yourself staring into the eyes of ginger-haired pigs; there were a few porkers and the odd one being fattened for bacon, but Mr. Worley's pride was his sows. He had six of them — Queenie, Princess, Ruby, Marigold, Delilah, and Primrose.

Nothing pleases Mr. Worley more than to sit down with Mr. Herriot and have "a piggy talk":

"You know, Mr. Herriot, sitting here talking like this with you, I'm 'appy as the king of England!"

On one occasion, at one o'clock in the morning,

Mr. Herriot comes to the rescue of Marigold, who is having trouble feeding her newborn piglets. Mr. Worley is distraught with worry, but Mr. Herriot once again solves the problem, and he celebrates his success in Mr. Worley's pub. Mr. Worley, however, choses to do his celebrating outside the pub:

The glow from the pig pen showed through the darkness of the yard, and as I crossed over, the soft rumble of pig and human voices told me that Mr. Worley was still talking things over with his sow. He looked up as I came in, and his face in the dim light was ecstatic.

"Mr. Herriot," he whispered. "Isn't that a beautiful sight?"

He pointed to the little pigs, who were lying motionless in a layered heap, sprawled over each other without plan or pattern, eyes tightly closed, stomachs bloated with Marigold's bountiful fluid.

"It is indeed," I said, prodding the sleeping mass with my finger but getting no response beyond the lazy opening of an eye. "You'd have to go a long way to beat it."

To Have and To Hold from This Day Forward

When I was a boy of six, my grandfather came to stay at our home in the country and brought with him his constant companion, a veteran tomcat named Tiger. Now, if Tiger had displayed any concern for the finer points of etiquette, I may never have come to appreciate the infinite qualities of pigs, but it so happens that Tiger committed the grave indiscretion of eating my pet canary for breakfast on the very first morning. In an effort to compensate, my grand-father gave me ten shillings to buy another bird; however, fearing that this might also fall victim to Tiger's table, I decided to look for a more durable pet.

The problem was quickly resolved by the advent of the local village flower show and fête, at which one of the side shows featured bowling for the pig at a shilling a turn. To my intense delight, after spending only two shillings, I won the prize, which turned out to be the runt of a litter of Wiltshire Whites, with one enormous black eye. I immediately named her Black-Eyed Susan, bore her home in triumph, and added her to the already extensive family of dogs, cats, rabbits, and geese.

In this environment, Susan flourished and soon grew from a bottle-fed runt to a lady of substantial proportions, consuming two gallons of milk a day. But for all her size, she insisted on taking her place as a full member of the family and would follow me for hours, walking to heel along with the dogs. She was a devoted and constant companion and bore with extreme good nature the many demands made upon her by the three of us children. She would double for bucking bronco or trap pony at will and was expert at shaking apples out of trees, which activity did not endear her to my father.

Susan hated being left behind and always had to be securely locked in the orchard when we went out on more formal occasions. One Sunday, she succeeded in breaking out and promptly traced us to church, where she precipitated a noisy reunion at the altar steps during the sermon. The vicar, being a compassionate man, eased my embarrassment and my mother's wrath by commenting from the pulpit that he was sure the good Lord was happy to see Susan in His house, which was more than could be said of many of his parishioners.

Although she was very gentle with us children and with members of the family, Susan was a formidable watchdog, and for the years she lived in our orchard, we were never robbed of any fruit. Indeed, on one occasion, when she had piglets, she treed the plumber and the carpenter on their lunch break and kept them captive with her screams of rage until we returned a full hour later.

It is fair to say that Susan was a character, and in her way, contributed a great deal to the happiness of those early years. She certainly laid the foundation for my enduring respect for animals and my special affection for her breed.

William Nicholls

In Sickness and in Health

Pig fever — not a pleasant thing to contract. It is an all-consuming, worrisome sickness that strikes a human whose pig is not on form (and it becomes worse when the Agricultural Show is just over a week away). In P. G. Wodehouse's story "Pig-hoo-o-o-o-ey!" Lord Emsworth, a gentleman farmer, is stricken when his prize pig, Empress of Blandings, refuses her food. Her hunger strike coincides exactly with the arrest of her keeper, George Cyril Wellbeloved, for disorderly behavior following a drunken binge at the local Goat and Feathers pub:

> On July the twentieth, Empress of Blandings, always hitherto a hearty and even a boisterous feeder, for the first time on record declined all nourishment. And on the morning of July the twenty-first, the veterinary surgeon called in to diagnose and deal with this strange asceticism, was compelled to confess to Lord Emsworth that the thing was beyond his professional skill....
>
> Let us just see, before proceeding, that we have got these dates correct:

July 18. — Birthday Orgy of Cyril Wellbeloved.

July 19. — Incarceration of Ditto.

July 20. — Pig Lays off the Vitamins.

July 21. — Veterinary Surgeon Baffled.

Right.

The effect of the veterinary surgeon's announcement on Lord Emsworth was overwhelming. As a rule, the wear and tear of our complex modern life left this vague and amiable peer unscathed. So long as he had sunshine, regular meals, and complete freedom from the society of his younger son, Frederick, he was placidly happy. But there were chinks in his armor, and one of these had been pierced this morning. Dazed by the news, he stood at the window of the great library of Blandings Castle, looking out with unseeing eyes.

Lord Emsworth's sister, Lady Constance, is only concerned that her niece, Angela, has forsaken her engagement to the blue-blooded Heacham for a "hopeless ne'er-do-well" by the name of James Belford:

It seemed to Lord Emsworth that there was a frightful amount of conversation going on. He had the sensation of hav-

ing become a mere bit of flotsam upon a tossing sea of female voices. Both his sister and his niece appeared to have much to say, and they were saying it simultaneously and *fortissimo*. He looked wistfully at the door.

It was smoothly done. A twist of the handle, and he was beyond those voices where there was peace. Galloping gaily down the stairs, he charged out into the sunshine.

His gaiety was not long lived. Free at last to concentrate itself on the really serious issues of life, his mind grew sombre and grim. Once more there descended upon him the cloud which had been oppressing his soul before all this Heacham-Angela-Belford business began. Each step that took him nearer the sty where the ailing Empress resided seemed a heavier step than the last. He reached the sty; and, draping himself over the rails, peered moodily at the vast expanse of pig within.

For, even though she had been doing a bit of dieting of late, Empress of Blandings was far from being an ill-nourished animal. She resembled a captive balloon with ears and a tail, and was as

nearly circular as a pig can be without bursting. Nevertheless, Lord Emsworth, as he regarded her, mourned and would not be comforted. A few more square meals under her belt, and no pig in all Shropshire could have held its head up in the Empress's presence. And now, just for the lack of those few meals, the supreme animal would probably be relegated to the mean obscurity of an "Honourably Mentioned." It was bitter, bitter.

But James Belford has more to offer than Lord Emsworth suspects:

"Are you employed on a farm?"

"I was employed on a farm."

"Pigs?" said Lord Emsworth in a low, eager voice.

"Among other things."

Lord Emsworth gulped. His fingers clutched at the tablecloth.

"Then perhaps, my dear fellow, you can give me some advice. For the last two days my prize sow, Empress of Blandings, has declined all nourishment. And the Agricultural Show is on Wednesday week. I am distracted with anxiety."

James Belford frowned thoughtfully.

"What does your pig-man say about it?"

"My pig-man was sent to prison two days ago. Two days!" For the first time the significance of the coincidence struck him. "You don't think that can have anything to do with the animal's loss of appetite?"

"Certainly. I imagine she is missing him and pining away because he isn't there."

Lord Emsworth was surprised. He had only a distant acquaintance with George Cyril Wellbeloved, but from what he had seen of him he had not credited him with this fatal allure.

"She probably misses his afternoon call."

Again his lordship found himself perplexed. He had had no notion that pigs were such sticklers for the formalities of social life.

"His call?"

"He must have had some special call that he used when he wanted her to come to dinner. One of the first things you learn on a farm is hog-calling. Pigs are temperamental. Omit to call them, and they'll starve rather than put on the nose-bag. Call them right, and they will follow you to the ends of the earth

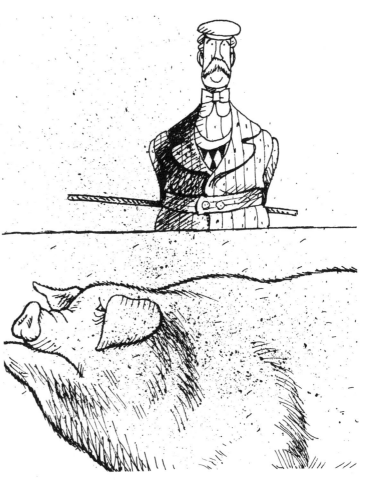

with their mouths watering."

"God bless my soul! Fancy that."

"A fact, I assure you. These calls vary in different parts of America. In Wisconsin, for example, the words, 'Poig, Poig, Poig' bring home — in both the literal and the figurative sense — the bacon. In Illinois, I believe they call 'Burp, Burp, Burp,' while in Iowa the

phrase 'Klus, Klus, Klus' is preferred. Proceeding to Minnesota, we find 'Peega, Peega, Peega' or, alternatively, 'Oink, Oink, Oink,' whereas in Milwaukee, so largely inhabited by those of German descent, you will hear the good old Teuton 'Komm Schweine, Komm Schweine.' Oh, yes, there are all sorts of pig-calls, from the Massachusetts 'Phew, Phew, Phew' to the

'Loo-ey, Loo-ey, Loo-ey' of Ohio, not counting various local devices such as beating on tin cans with axes or rattling pebbles in a suitcase. I knew a man out in Nebraska who used to call his pigs by tapping on the edge of the trough with his wooden leg."

"Did he, indeed?"

"But a most unfortunate thing happened. One evening, hearing a woodpecker at the top of a tree, they started shinning up it; and when the man came out, he found them all lying there in a circle with their necks broken."

"This is no time for joking," said Lord Emsworth, pained.

"I'm not joking. Solid fact. Ask anybody out there."

Lord Emsworth placed a hand to his throbbing forehead.

"But if there is this wide variety, we have no means of knowing which call Wellbeloved . . . ."

"Ah," said James Belford, "but wait. I haven't told you all. There is a masterword."

"A what?"

"Most people don't know it, but I had it straight from the lips of Fred Patzel, the hog-calling champion of the Western States. What a

man! I've known him to bring pork chops leaping from their plates. He informed me that, no matter whether an animal has been trained to answer the Illinois 'Burp' or the Minnesota 'Oink,' it will always give immediate service in response to this magic combination of syllables. It is to the pig world what the Masonic grip is to the human. 'Oink' in Illinois or 'Burp' in Minnesota, and the animal merely raises its eyebrows and stares coldly. But go to either State and call 'Pig-hoo-oo-ey!' . . . . "

The expression on Lord Emsworth's face was that of a drowning man who sees a lifeline.

"Is that the masterword of which you spoke?"

"That's it."

"Pig — ?"

" — hoo-oo-ey."

"Pig-hoo-o-ey?"

"You haven't got it right. The first syllable should be short and staccato, the second long and rising into a falsetto, high but true."

"Pig-hoo-o-o-ey."

"Pig-hoo-o-o-ey."

"Pig-hoo-o-o-ey!" yodeled Lord Emsworth, flinging his head back and giving tongue in a high, penetrating tenor which caused ninety-three

Senior Conservatives, lunching in the vicinity, to congeal into living statues of alarm and disapproval.

"More body to the 'hoo'," advised James Belford.

"Pig-hoo-o-o-o-ey!"

The senior Conservative Club is one of the few places in London where lunchers are not accustomed to getting music with their meals. White-whiskered financiers gazed bleakly at bald-headed politicians, as if asking silently what was to be done about this. Bald-headed politicians stared back at white-whiskered financiers, replying in the language of the eye that they did not know. The general sentiment prevailing was a vague determination to write to the Committee about it.

"Pig-hoo-o-o-ey!" carroled Lord Emsworth. And, as he did so, his eye fell on the clock over the mantelpiece. Its hands pointed to twenty minutes to two.

He started convulsively. The best train in the day for Market Blandings was the one which left Paddington station at two sharp. After that there was nothing till the five-five.

He was not a man who

often thought; but, when he did, to think was with him an act. A moment later he was scudding over the carpet, making for the door that led to the broad staircase.

Throughout the room which he had left, the decision to write in strong terms to the Committee was now universal; but from the mind, such as it was, of Lord Emsworth the past, with the single exception of the word "Pig-hoo-o-o-o-ey!", had been completely blotted.

The time comes for Lord Emsworth, who has perfected his "pig call," to gage its results. Along with his butler and niece, he ventures forward:

The bijou residence of the Empress of Blandings looked snug and attractive in the moonlight. But beneath even the beautiful things of life, there is always an underlying sadness. This was supplied in the present instance by a long, low trough, only too plainly full to the brim of succu-

lent mash and acorns. The fast, obviously, was still in progress.

The sty stood some considerable distance from the castle walls, so that there had been ample opportunity for Lord Emsworth to rehearse his little company during the journey. By the time they had ranged themselves against the rails, his two assistants were letter-perfect.

"Now," said his lordship.

There floated out upon the summer's night a strange composite sound that sent the birds roosting in the trees above shooting off their perches like rockets. Angela's clear soprano rang out like the voice of the village blacksmith's daughter. Lord Emsworth contributed a reedy tenor. And the bass notes of Beach probably did more to startle the birds than any other one item in the program.

They paused and listened. Inside the Empress's boudoir, there sounded the movement of a heavy body. There was an inquiring grunt. The next moment the sacking that covered the doorway was pushed aside, and the noble animal appeared.

"Now!" said Lord Emsworth again.

Once more that musical cry shattered the silence of the night. But it brought no responsive movement from Empress of Blandings. She stood there motionless, her nose elevated, her ears hanging down, her eyes everywhere but on the trough where, by rights, she should now be digging in and getting hers. A chill disappointment crept over Lord Emsworth, to be succeeded by a gust of petulant anger.

"I might have known it," he said bitterly. "That young scoundrel was deceiving me. He was playing a joke on me."

"He wasn't," cried Angela indignantly. "Was he, Beach?"

"Not knowing the circumstances, Miss, I cannot venture an opinion."

"Well, why has it no effect, then?" demanded Lord Emsworth.

"You can't expect it to work right away. We've got her stirred up, haven't we? She's thinking it over, isn't she? Once more will do the trick. Ready, Beach?"

"Quite ready, Miss."

"Then when I say three. And this time, Uncle Clarence, do please for goodness' sake not yowl like you did before. It was enough to put any pig off. Let it come out quite easily and gracefully. Now, then. One, two — three!"

The echoes died away. And as they did so a voice spoke.

"Community singing?"

"Jimmy!" cried Angela, whisking round.

"Hullo, Angela. Hullo, Lord Emsworth. Hullo, Beach."

"Good evening, sir. Happy to see you once more."

"Thanks. I'm spending a few days at the Vicarage with my father. I got down here by the five-five."

Lord Emsworth cut peevishly in upon these civilities.

"Young man," he said, "what do you mean by telling me that my pig would respond to that cry? It does nothing of the kind."

"You can't have done it right."

"I did it precisely as you instructed me. I have had, moreover, the assistance of Beach here and my niece Angela — "

"Let's hear a sample."

Lord Emsworth cleared his throat. "Pig-hoo-o-o-o-ey!"

James Belford shook his head.

"Nothing like it," he said. "You want to begin the 'Hoo' in a low minor of two quarter notes in four-four time. From this build gradually to a higher note, until at last the voice is soaring in full crescendo, reaching F sharp on the natural scale and dwelling for two retarded half-notes, then breaking into a shower of accidental grace-notes."

"God bless my soul!" said Lord Emsworth, appalled. "I shall never be able to do it."

"Jimmy will do it for you," said Angela. "Now that he's engaged to me, he'll be one of the family and always popping about here. He can do it every day till the show is over."

James Belford nodded.

"I think that would be the wisest plan. It is doubtful if an amateur could ever produce real results. You need a voice that has been trained on the open prairie and that has gathered richness and strength from competing with tornadoes. You need a manly, wind-scorched voice with a suggestion in it of the crackling of corn husks and the whisper of evening breezes in the fodder. Like this!"

Resting his hands on the rail before him, James

Belford swelled before their eyes like a young balloon. The muscles on his cheekbones stood out, his forehead became corrugated, his ears seemed to shimmer. Then, at the very height of the tension, he let it go like, as the poet beautifully puts it, the sound of a great Amen.

"PIG-HOOOOO-OOO-OOO-O-O-ey!"

They looked at him, awed. Slowly, fading off across hill and dale the vast bellow died away. And suddenly, as it died, another, softer sound succeeded it. A sort of gulpy, gurgly, plobby, squishy, woffle-some sound, like a thousand eager men drinking soup in a foreign restaurant. And, as he heard it, Lord Emsworth uttered a cry of rapture.

The Empress was feeding.

Till Death Us Do Part

Death of a Pig

Autumn 1947

I spent several days and nights in mid-September with an ailing pig, and I feel driven to account for this stretch of time, more particularly since the pig died at last, and I lived, and things might easily have gone the other way round and none left to do the accounting. Even now, so close to the event, I cannot recall the hours sharply and am not ready to say whether death came on the third night or the fourth night. This uncertainty afflicts me with a sense of personal deterioration; if I were in decent health, I would know how many nights I had sat up with a pig.

The scheme of buying a spring pig in blossomtime, feeding it through summer and fall, and butchering it when the solid cold weather arrives, is a familiar scheme to me and follows an antique pattern. It is a tragedy enacted on most farms with perfect fidelity to the original script. The murder, being premeditated, is in the first degree but is quick and skillful, and the smoked bacon and ham provide a ceremonial ending whose fitness is seldom questioned.

Once in a while something slips — one of the actors goes up in his lines and the whole performance stumbles and halts. My pig simply failed to show up for a meal. The alarm spread rapidly. The classic outline of the tragedy was lost. I found myself cast suddenly in the role of pig's friend and physician — a farcical character with an enema bag for a prop. I had a presentiment, the very first afternoon, that the play would never regain its balance and that my sympathies were now wholly with the pig. This was slapstick — the sort of dramatic treatment that instantly appealed to my old dachshund, Fred, who joined the vigil, held the bag, and, when all was over, presided at the interment. When we slid the body into the grave, we both were shaken to the core. The loss we felt was not the loss of ham but the loss of pig. He had evidently become precious to me, not that he represented a distant nourishment in a hungry time, but that he had suffered in a suffering world. But I'm running ahead of my story and shall have to go back.

My pigpen is at the bottom of an old orchard below the house. The pigs I have raised have lived in a faded building that once was an icehouse. There is a pleasant yard to move about in, shaded by an apple tree that overhangs the low rail fence. A pig couldn't ask for anything better — or none has, at any rate. The sawdust in the icehouse makes a comfortable bottom in which to root, and a warm bed. This sawdust, however, came under suspicion when the pig took sick. One of my neighbors said he thought the pig would have done better on new ground — the same principle that applies in planting potatoes. He said there might be something unhealthy about that sawdust, that he never thought well of sawdust.

It was about four o'clock in the afternoon when I first noticed that there was something wrong with the pig. He failed to appear at the trough for his supper, and

when a pig (or a child) refuses supper, a chill wave of fear runs through any household, or ice-household. After examining my pig, who was stretched out in the sawdust inside the building, I went to the phone and cranked it four times. Mr. Dameron answered. "What's good for a sick pig?" I asked. (There is never any identification needed on a country phone; the person on the other end knows who is talking by the sound of the voice and by the character of the question.)

"I don't know, I never had a sick pig," said Mr. Dameron, "but I can find out quick enough. You hang up and I'll call Henry."

Mr. Dameron was back on the line again in five minutes. "Henry says roll him over on his back and give him two ounces of castor oil or sweet oil, and if that doesn't do the trick give him an injec-

tion of soapy water. He says he's almost sure the pig's plugged up, and even if he's wrong, it can't do any harm."

I thanked Mr. Dameron. I didn't go right down to the pig, though. I sank into a chair and sat still for a few minutes to think about my troubles, and then I got up and went to the barn, catching up on some odds and ends that needed tending to. Unconsciously I held off, for an hour, the deed by which I would officially recognize the collapse of the performance of raising a pig; I wanted no interruption in the regularity of feeding, the steadiness of growth, the even succession of days. I wanted no interruption, wanted no oil, no deviation. I just wanted to keep on raising a pig, full meal after full meal, spring into summer into fall. I didn't even know whether there were two ounces of castor oil on

the place.

Shortly after five o'clock, I remembered that we had been invited out to dinner that night and realized that if I were to dose a pig, there was no time to lose. The dinner date seemed a familiar conflict: I move in a desultory society, and often a week or two will roll by without my going to anybody's house to dinner or anyone's coming to mine, but when an occasion does arise, and I am summoned, something usually turns up (an hour or two in advance) to make all human intercourse seem vastly inappropriate. I have come to believe that there is in hostesses a special power of divination, and that they deliberately arrange dinners to coincide with pig failure or some other sort of failure. At any rate, it was after five o'clock and I knew I could put off no longer the evil hour.

When my son and I arrived at the pigyard, armed with a small bottle of castor oil and a length of clothesline, the pig had emerged from his house and was standing in the middle of his yard, listlessly. He gave us a slim greeting. I could see that he felt uncomfortable and uncertain. I had brought the clothesline thinking I'd have to tie him (the pig weighed more than 100 pounds) but we never used it. My son reached down, grabbed both front legs, upset him quickly, and when he opened his mouth to scream, I turned the oil into his throat — a pink, corrugated area I had never seen before. I had just time to read the label while the neck of the bottle was in his mouth. It said Puretest. The screams, slightly muffled by oil, were pitched in the hysterically high range of pig-sound, as though torture were being carried out, but they didn't last long: it was all over rather suddenly, and, his legs released, the pig righted himself.

In the upset position, the corners of his mouth had been turned down, giving him a frowning expression. Back on his feet again, he regained the set smile that a pig wears even in sickness. He stood his ground, sucking slightly at the residue of oil; a few drops leaked out of his lips while his wicked eyes, shaded by their coy little lashes, turned on me in disgust and hatred. I scratched him gently with oily fingers and he remained quiet, as though trying to recall the satisfaction of being scratched when in health, and seeming to rehearse in his mind the indignity to which he had just been subjected. I noticed, as I stood there, four or five small dark spots on his back near the tail end, reddish-brown in color, each about the size of a housefly. I could not make out what they were. They did not look troublesome, but at the same time they did not look like mere surface bruises or chafe marks. Rather they seemed

blemishes of internal origin. His stiff white bristles almost completely hid them, and I had to part the bristles with my fingers to get a good look.

Several hours later, a few minutes before midnight, having dined well and at someone else's expense, I returned to the pighouse with a flashlight. The patient was asleep. Kneeling, I felt his ears (as you might put your hand on the forehead of a child) and they seemed cool, and then with the light made a careful examination of the yard and the house for sign that the oil had worked. I found none and went to bed.

We had been having an unseasonable spell of weather — hot, close days, with the fog shutting in every night, scaling for a few hours in midday, then creeping back again at dark, drifting in first over the trees on the point, then suddenly blowing across the fields, blotting out the world and taking possession of houses, men, and animals. Everyone kept hoping for a break, but the break failed to come. Next day was another hot one. I visited the pig before breakfast and tried to tempt him with a little milk in his trough. He just stared at it, while I made a sucking sound through my teeth to remind him of past pleasures of the feast. With very small, timid pigs, weanlings, this ruse is often quite successful and will encourage them to eat; but with a large, sick pig the ruse is senseless, and the sound I made must have made him feel, if anything, more miserable. He not only did not crave food, he felt a positive revulsion to it. I found a place under the apple tree where he had vomited in the night.

At this point, although a depression had settled over me, I didn't suppose that I was going to lose my pig. From the lustiness of a healthy pig a man derives a feeling of personal lustiness; the stuff that goes into the trough and is received with such enthusiasm is an earnest of some later feast of his own, and when this suddenly comes to an end and the food lies stale and untouched, souring in the sun, the pig's imbalance becomes the man's, vicariously, and life seems insecure, displaced, transitory.

As my own spirits declined, along with the pig's, the spirits of my vile old dachshund rose. The frequency of our trips down the footpath through the orchard to the pigyard delighted him, although he suffers greatly from arthritis, moves with difficulty, and would be bedridden if he could find anyone willing to serve him meals on a tray.

He never missed a chance to visit the pig with me, and he made many professional calls on his own. You could see him down there at all hours, his white face parting the grass along the fence as he wobbled and stumbled about, his stethoscope dangling — a happy quack, writing his villainous prescriptions and grinning his corrosive grin. When the enema bag appeared, and the bucket of warm suds, his happiness was complete, and he managed to squeeze his enormous body between the two lowest rails of the yard and then assumed full charge of the irrigation. Once, when I lowered the bag to check the flow, he reached in and hurriedly drank a few mouthfuls of the suds to test their potency. I have noticed that Fred will feverishly consume any substance that is associated with trouble — the bitter flavor is to his liking. When the bag was above reach, he concentrated on the pig and was everywhere at once, a tower of strength and inconvenience. The pig, curiously enough, stood rather quietly through this colonic carnival, and the enema, though ineffective, was not as difficult as I had anticipated.

I discovered, though, that once having given a pig an enema there is no turning back, no chance of resuming one of life's more stereotyped roles. The pig's lot and mine were inextricably bound now, as though the rubber tube were the silver cord. From then until the time of his death I held the pig steadily in the bowl of my mind; the task of trying to deliver him from his misery became a strong obsession. His suffering soon became the embodiment of all earthly wretchedness. Along toward the end of the afternoon, defeated in physicking, I phoned the veterinary twenty miles away and placed the case formally in his hands. He was full of questions, and when I casually mentioned the dark spots on the pig's back, his voice changed its tone.

"I don't want to scare you," he said, "but when there are spots, erysipelas has to be considered."

Together we considered erysipelas, with frequent interruptions from the telephone operator, who wasn't sure the connection had been established.

"If a pig has erysipelas can he give it to a person?" I asked.

"Yes, he can," replied the vet.

"Have they answered?" asked the operator.

"Yes, they have," I said. Then I addressed the vet again. "You better come over here and examine this pig right away."

"I can't come myself," said the vet, "but McFarland can come this evening if that's all right. Mac knows more about pigs than I do anyway. You needn't worry too much about the spots. To indicate erysipelas they would have to be deep hemorrhagic infarcts."

"Deep hemmorrhagic what?" I asked.

"Infarcts," said the vet.

"Have they answered?" asked the operator.

"Well," I said. "I don't know what you'd call these spots, except they're about the size of a housefly. If the pig has erysipelas I guess I have it, too, by this time, because

we've been very close lately."

"McFarland will be over," said the vet.

I hung up. My throat felt dry and I went to the cupboard and got a bottle of whiskey. Deep hemorrhagic infarcts — the phrase began fastening its hooks in my head. I had assumed that there could be nothing much wrong with a pig during the months it was being groomed for murder; my confidence in the essential health and endurance of pigs had been strong and deep, particularly in the health of pigs that belonged to me and that were part of my proud scheme. The awakening had been violent, and I minded it all the more because I knew that what could be true of my pig could be true also of the rest of my tidy world. I tried to put this distasteful idea from me, but it kept recurring. I took a short drink of the whiskey and then, although I wanted to go down to the yard and look for fresh signs, I was scared to. I was certain I had erysipelas.

It was long after dark and the supper dishes had been put away when a car drove in and McFarland got out. He had a girl with him. I could just make her out in the darkness — she seemed young and pretty. "This is Miss Owen," he said. "We've been having a picnic supper on the shore, that's why I'm late."

McFarland stood in the driveway and stripped off his jacket, then his shirt. His stocky arms and capable hands showed up in my flashlight's gleam as I helped him find his coverall and get zipped up. The rear seat of his car contained an astonishing amount of paraphernalia, which he soon overhauled, selecting a chain, a syringe, a bottle of oil, a rubber tube, and some other things I couldn't identify. Miss Owen said she'd go along with us and see the pig. I led the way down the warm slope of the orchard, my light picking out the path for them, and we all three climbed the fence, entered the pighouse, and squatted by the pig while McFarland took a rectal reading. My flashlight picked up the glitter of an engagement ring on the girl's hand.

"No elevation," said McFarland, twisting the thermometer in the light. "You needn't worry about erysipelas." He ran his hand slowly over the pig's stomach and at one point the pig cried out in pain.

"Poor piggledy-

wiggledy!" said Miss Owen.

The treatment I had been giving the pig for two days was then repeated, somewhat more expertly, by the doctor, Miss Owen and I handing him things as he needed them — holding the chain that he had looped around the pig's upper jaw, holding the syringe, holding the bottle stopper, the end of the tube, all of us working in darkness and in comfort, working with the instinctive teamwork induced by emergency conditions, the pig unprotesting, the house shadowy, protecting, intimate. I went to bed tired but with a feeling of relief that I had turned over part of the responsibility of the case to a licensed

doctor. I was beginning to think, though, that the pig was not going to live.

He died twenty-four hours later, or it might have been forty-eight — there is a blur in time here, and I may have lost or picked up a day in the telling and the pig one in the dying. At intervals during the last day I took cool fresh water down to him, and at such times as he found the strength to get to his feet he would stand with head in the pail and snuffle his snout around. He drank a few sips but no more; yet it seemed to comfort him to dip his nose in water and bobble it about, sucking in and blowing out through his teeth. Much of the time, now, he lay indoors half buried in sawdust. Once, near the last, while I was attending him I saw him try to make a bed for himself but he lacked the strength, and when he set his snout into the dust he was unable to plow even the little furrow he needed to lie down in.

He came out of the house to die. When I went down, before going to bed, he lay stretched in the yard a few feet from the door. I knelt, saw that he was dead, and left him there: his face had a mild look, expressive neither of deep peace nor of deep suffering, although I think he had suffered a good deal. I went back up to the house and to bed, and cried internally — deep hemorrhagic intears. I didn't wake till nearly eight the next morning, and when I looked out the open window the grave was already being dug, down beyond the dump under a wild apple. I could hear the spade strike against the small rocks that blocked the way. Never send to know for whom the grave is dug, I said to myself, it's dug for thee. Fred, I well knew, was supervising the work of digging, so I ate breakfast slowly.

It was a Saturday morning. The thicket in which I found the gravediggers at work was dark and warm, the sky overcast. Here, among alders and young hackmatacks, at the foot of the apple tree, Lennie had dug a beautiful hole, five feet long, three feet wide, three feet deep. He was standing in it, removing the last spadefuls of earth while Fred patrolled the brink in simple but impressive circles, disturbing the loose earth of the mound so that it trickled back in. There had been no rain in weeks and the soil, even three feet down, was dry and powdery. As I stood and stared, an enormous earthworm, which had been partially exposed by the spade at the bottom, dug itself deeper and made a slow withdrawal, seeking even remoter moistures at even lonelier depths. And just as Lennie stepped out and rested his spade against the tree and lit a cigarette, a small green apple separated itself from a branch overhead and fell into the hole. Everything about this last scene seemed overwritten — the dismal sky, the shabby woods, the imminence of rain, the worm (legendary bedfellow of the dead), the apple (conventional garnish of a pig.)

But even so, there was a directness and dispatch about animal burial, I thought, that made it a more decent affair than human burial: there was no stopover in the undertaker's foul parlor, no wreath nor spray; and when we hitched a line to the pig's hind legs and dragged him swiftly from his yard, throwing our weight into the harness and leaving a wake of crushed grass and smoothed rubble over the dump, ours was a businesslike procession, with Fred, the dishonorable pallbearer, staggering along in the rear, his perverse bereavement showing in every seam in his face; and the postmortem performed handily and swiftly right at the edge of the grave, so that the inwards that had caused the pig's death preceded him into the ground, and he lay at last resting squarely on the cause of his own undoing.

I threw in the first shovelful, and then we worked rapidly and without talk, until the job was complete. I picked up the rope, made it fast to Fred's collar (he is a notorious ghoul), and we all three filed back up the path to the house, Fred bringing up the rear and holding back every inch of the way, feigning unusual stiffness. I noticed that although he weighed far less than the pig, he was harder to drag, being possessed of the vital spark.

The news of the death of my pig traveled fast and far, and I received many expressions of sympathy from friends and neighbors, for no one took the event lightly, and the premature expiration of a pig is, I soon discovered, a departure that the community marks solemnly on its calendar, a sorrow in which it feels fully involved. I have written this account in penitence and in grief, as a man who failed to raise his pig, and to explain my deviation from the classic course of so many raised pigs. The grave in the woods is unmarked, but Fred can direct the mourner to it unerringly and with immense good will, and I know he and I shall often revisit it, singly and together, in seasons of reflection and despair, on flagless memorial days of our own choosing.

E. B. White

Let No Man Put Asunder

Pigs are not easily pushed around. They will not budge when bullied and are not fooled by sweet talk. Pigs are not disdainful like cats, nor do they cringe like dogs — they look you straight in the eye, without so much as batting a bristly eyelid.

Patricia Highsmith's story "In the Dead of Truffle Season" tells of one pig who is pushed beyond the limits of endurance. His name is Samson. Samson belongs to Emile and lives in the Lot region of France. In winter, Emile takes Samson truffle hunting along with his friend René, René's dog, and, of course, the inevitable bottle of Armagnac. Emile is proud of Samson, who is a truffle hunter of some repute even though he is slow to root up the delicacies because he never gets a chance to eat them; his only reward is a piece of cheese. Samson resents this poor payment. He also dislikes the way Emile kicks and prods him to hurry, and he hates the local dogs, which he keeps at bay with a snort and toss of his massive head. Luckily for them, his tusks have been sawn off. Nevertheless, he remains a formidable brute. Even the barnyard pigs treat him with respect, relinquishing the trough when he approaches.

In January, it is announced that a pâté manufacturer is holding a truffle-hunting contest and offering three valuable prizes, including a cuckoo clock and a transistor radio, for the winners. There is tremendous local excitement. The village talks of nothing else, especially as the contest is to be covered by the newspapers and even the television. Emile will take Samson, and René will hunt with his pig, Lunache. The other men plan to use dogs. Emile dreams of the cuckoo clock and of being interviewed on television. He feels sure that he will win, as Samson is the best local truffle hunter. The big day arrives. All the men gather in the village with their sacks and dogs or pigs. At the start, Emile and another man, François, head off in the same direction, where each knows a lot of truffles are growing. Sure enough, they find an area with plenty of truffles for both. Emile is impatient — there is not a moment to lose. He prods and kicks Samson, who is just begin-

ning to smell the truffles. Their smell is too much for the pig. He drools and strains at the rope that restrains him as Emile unearths the goodies. Suddenly the rope breaks, and Samson immediately starts gobbling up the truffles. A furious, cursing Emile beats him and manages to tie him up again. But Samson has had enough. He breaks free once more, knocks Emile down, and rushes over to François's truffles, which he begins to devour. François is livid and, being jealous of Emile anyway, threatens to have him disqualified from the contest if he doesn't control his pig. François sets out toward the village to fulfill his threat. Emile, in desperation, tries to appease Samson with some cheese, but again the smell of truffles in the sack wafts through the air, and again Samson starts drooling. And then Samson charges:

Somehow the pig's belly hit Emile in the face, or the point of his chin, and Emile was knocked half unconscious. He shook his head and made sure he still had a good grip on his fork. He had suddenly realized that Samson could and might kill him, if he didn't protect himself.

"Au secours!" Emile

yelled. *"Help!"*

Emile brandished the fork at Samson, intending to scare the pig off while he got to his feet.

Samson had no intention, except to protect himself. He saw the fork as an enemy, a very clear challenge, and he blindly attacked it. The fork went askew and dropped as if limp. Samson's front hooves stood triumphant on Emile's abdomen. Samson snooted. And Emile gasped, but only a few times.

The awful pink and damp nose of the pig was almost in Emile's face, and he recalled from childhood many pigs he had known, pigs who had seemed to him as gigantic as this Samson now crushing the breath out of him. Pigs, sows, piglets of all patterns and coloring seemed to combine and become this one monstrous Samson who most certainly — Emile now knew it — was going to kill him, just by standing on him. The fork was out of reach. Emile flailed his arms with his last strength, but the pig wouldn't budge. And Emile could not gasp one breath of air. Not even an animal any longer, Emile thought, this pig, but an awful, evil force in a most hideous form. Those tiny, stupid eyes in the grotesque flesh! Emile tried to call out and found that he couldn't make as much noise as a small bird.

When the man became quiet, Samson stepped off his body and nuzzled him in the side to get at the truffle sack again. Samson was calming down a bit. He no longer held his breath or panted, as he had done alternately for the last minutes, but began to breathe normally. The heavenly scent of truffles further soothed him. He snuffled, sighed, inhaled, ate, his snout and tongue seeking out the last morsels from the corners of the khaki sack. And all his own gleanings! But this thought came not at all clearly to Samson. In fact, he had a vague feeling that he was going to be shooed away from his banquet, yet who was there to shoo him away now? This very special sack, into which he had seen so many black truffles vanishing, out of which had come measly, contemptible crumbs of yellow cheese — all that was finished, and now the sack was his. Samson even ate some of the cloth.

The following day, Samson finds a farm and a new owner called Alphonse, who debates whether to send Samson to market or to try him at truffling:

Samson grew a little fatter, and dominated the other pigs, two sows and their piglets. The food was slightly different and more abundant than at the other farm. Then came the day — an ordinary working day, it seemed to Samson from the look of the farm — when he was taken on a lead to go to the woods for truffles. Samson trotted along in good spirits. He intended to eat a few truffles today, besides finding them for the man. Somewhere in his brain, Samson was already thinking that he must from the start show this man that he was not to be bossed.

Dogs look up at you,
Cat's look down on you,
But pigs is equal.

Old English saying

Pig Pourri

The Anatomy of a Pig

Pigs and human beings have similar digestive systems, teeth, and blood. Pigs' skin is used as a substitute for ours in the treatment of burns, and pig-produced secretions are used for human medication. Even the flavor of our flesh is apparently the same. In cannibalistic cultures, human meat is called "long pig" because it resembles the taste and look of pork, although human bones are obviously longer. Recently, it has been discovered that pig waste can be used as an alternative source of energy. Pig manure, when it decomposes, releases quantities of methane large enough to fuel any heating system.

Pigs on Coins

The Bermuda Cent:

Still used today, a brass coin bearing a wild hog was first introduced to Bermuda in the early seventeenth century. Wild hogs were regularly spotted by seafarers visiting the isle, which was aptly known as "the earthly paradise."

The Irish Halfpenny:

This copper coin, which was circulated as late as the mid-1960s, featured a sow with her young. A connection could be made here with the age-old practice whereby farmers paid their rents with animals, and pigs were always popular currency.

The Roman Coin:

In the latter half of the fourth century B.C., the pig adorned coins issued as souvenirs of the Festival of the Thesmophoria. During this festival, pigs were flung over a chasm as an offering, and later their rotting remains served as fertilizer for corn seed. White cakes, made out of flour and gypsum and served at the festival, were baked in the shape of pigs.

Crafted Pigs

Porcelain:

The word "porcelain" is derived from the Italian word "porcella," meaning "young sow." There is a type of sea shell, also called porcelain, which is smooth and white, not dissimilar, in fact, to the look and feel of a sow's back.

Piggy Banks:

Traditionally of a china mould, these ornaments are used for holding and saving loose change. The pig shape may have some association with the animal's unfair reputation for being greedy and miserly.

Pig Mugs:

In the late eighteenth and nineteenth centuries in England, clay, pig-shaped drinking vessels were all the rage, bringing recognition to the Sussex and Kentish pottery pieces. The head was detachable and formed a cup, which rested rather precariously on the snout. This mug is still in production, and in the more recent models, the ears of the pig have been lengthened to enable the cup to stand on a firmer base.

The Pig as Mascot:

In China, traders believe that pigs of gold and silk bring good fortune. And in Ireland, the figure of a pig is also considered lucky, but only if some part of it is broken.

Sign Language

The signs of public houses bear many piggy resemblances:

The Pig and Whistle:

In Scotland, "pig" means pot or mug, and "whistle" refers to small change. (Perhaps the saying "going to pot" came from here.)

Pig and Tinderbox:

This pub was originally called the "Elephant and Castle," but the picture looked more like an elephantine pig standing on a tinderbox, and so the deviation.

La Pique et le Carreau:

The cockney dialect corrupted the name to "The Pig and the Carrot," although the literal translation is "The Spade and the Diamond."

The Dorking Beacon:

... became "The Dog and Bacon." And why not, seeing that cockneys referred to beer as "the pig's ear"?

Other Deviations

Hog:

Some contend that the word "hog" comes from the Hebrew word "choog," meaning "to encompass or surround," suggestive of a round, fat figure. Others say it is a derivative of an Arabic phrase meaning "to have narrow eyes."

Casting Pearls Before Swine:

Although the source of this famous saying is unknown, T.H. White muses in his *Book of Beasts* that scholars of the Middle Ages managed to mix up the translation of "margarite" (pearl) with "marguerite" (flower). Perhaps this phrase was misconstrued and actually meant "feeding the swine on daisies"!

Das Kann Kein Schwein Lesen:

This German phrase is loosely translated as "no Pig can read this" (because it is illegible), and is rooted in a seventeenth century anecdote about a Saxon family called Swyn, who were scholars always ready to help illiterate peasants decipher letters and documents. Not all the handwriting presented to the Swyns was readable, however, and when even they were stumped, it was commented that "no Swyn can read this."

Pig Latin:

And for all those really wanting to be misunderstood — Ig-Pay-Atin-Lay.

Pigs in the Stars:

Chinese astrology has a bestial base. The months and years of "The Pig" are as follows:

January 30, 1911 to
February 18, 1912;
February 16, 1923 to
February 5, 1924;
February 4, 1935 to
January 24, 1936;
January 22, 1947 to
February 10, 1948;
February 8, 1959 to
January 28, 1960;
January 27, 1971 to
February 14, 1972;
February 13, 1983 to
February 1, 1984;
January 31, 1995 to
February 18, 1996.

Acknowledgements

The authors wish to thank the following sources for kind permission to include textual excerpts in this book: Abelard and Schumann, Harper & Row Publishers Inc. for the excerpt from *The Witchcraft World*, by G.L. Simons, 1974, pages 50, 51; Atheneum Publishers for the excerpt from "What is her name?... grumbled Garth" from *Mr. and Mrs. Pig's Evening Out*, written and illustrated by Mary Rayner, © 1976 Mary Rayner, page 86, and for the excerpt from "As I Looked Out" from *This Little Pig-A-Wig*, chosen by Lenore Blegvad and illustrated by Erik Blegvad, © 1978 Erik Blegvad, page 89; The Bermuda Dept. of Tourism, page 141; Jonathan Cape publishers for the excerpt from *The Book of Beasts*, by T.H. White, page 79; Chappell Music Limited for "The Warthog Song," words and music by Donald Swann and Michael Flanders, © 1955 Chappell & Co. Ltd., London, pages 112, 113; William Collins Sons and Company Limited, London for the excerpts from *Collins Concise Encyclopedia of Greek and Roman Mythology*, by Sabine G. Oswalt, © 1969, pages 41, 43; Coward, McCann & Geoghegan Inc. for the excerpts from *Lord of the Flies*, by William Golding, © 1954 William Golding, pages 51-54; Dash Music Co. Ltd. for the lyrics from "The Pig Got Up And Slowly Walked Away," by Benj. H. Burt, English version by Terry Sullivan, reproduced by permission of Dash Music Co. Ltd., 37 Soho Square, London W.1. for the British Empire (ex. Canada, Newfoundland and Australasia), and the continent of Europe, page 77; J.M. Dent and The Trustees for the Copyright of the late Dylan Thomas for the excerpt from *Under Milk Wood*, by Dylan Thomas, page 119; Doubleday & Company, Inc. for the excerpt from "Nor is there any beast..." from *The Hog Book*, © 1978 William B. Hedgepeth, page 31; E.P. Dutton & Company, Inc. for the excerpt from "Pooh and Piglet Go Hunting and Nearly Catch a Woozle" from *Winnie-the-Pooh*, by A.A. Milne, © 1926 E.P. Dutton & Company, Inc., renewal, 1954 by A.A. Milne, page 105; Faber & Faber Limited for the excerpts from *Lord of the Flies*, by William Golding, pages 51-54; Barthold Fles, Literary Agent for a description of the "Boar's Head Carol" from *A Treasury of Christmas Songs and Carols*, edited and annotated by Henry W. Wilson, 1973, page 47; *Fugue Magazine*, Toronto for the excerpt from *Musical World* (November 1839), page 15; Thomas Hagey and *Playboar Magazine* for the excerpt from *Playboar Magazine*, page 63; Hamish Hamilton Limited for *This Little Pig-A-Wig*, chosen by Lenore Blegvad and illustrated by Erik Blegvad, page 89, and for the excerpts from *Charlotte's Web*, by E. B. White, © E. B. White, pages 100-104; Harper & Row Publishers Inc. for the excerpt from *Small Pig*, written and illustrated by Arnold Lobel, © 1969 Arnold Lobel, page 88, for the excerpts from *Charlotte's Web*, by E.B. White, © 1952 E.B. White, pages 100-104; and for "Death of a Pig" from *Essays of E.B. White*, by E.B. White, © 1947 E.B. White, pages 130-137; F.B. Haviland Pub. Co. for the lyrics from "The Pig Got Up and Slowly Walked Away," by Benj. H. Burt, © F.B. Haviland Pub. Co., 58 W. 45 St., New York, N.Y., 10036, reprinted with permission of Jerry Vogel Music Co., Inc., page 77; William Heinemann Ltd., Publishers for the excerpt from *Pig Tale* by Helen Oxenbury, page 88, and for the excerpts from "In the Dead of Truffle Season" from *The Animal-Lover's Book of Beastly Murder*, by Patricia Highsmith, pages 138-140; Patricia Highsmith for the excerpts from "In the Dead of Truffle Season," which was first published by William Heinemann, London in 1975 in Patricia Highsmith's short story collection *The Animal-Lover's Book of Beastly Murder*, © 1975 Patricia Highsmith, pages 138-140; Hutchinson Publishers for the excerpts from *Hunting Weapons—The Arms and Armour Series*, by H.L. Blackmore, © 1971, first published by Barrie & Jenkins, pages 24, 30, 31, ("Henry VIII was a great..."), 32; Michael Joseph Ltd. for the excerpts from *All Creatures Great and Small*, by James Herriot, pages 120; Joanne Kates for the excerpt from "Truffles, An Underground Delicacy," page 20; Llewellyn Publications for the excerpt from *Witchcraft from the Inside*, by Raymond Buckland, page 46; Macmillan, London and Basingstoke for the excerpts from *Modern Pig-Sticking*, by A.E. Wardrop, pages 34, 36-38, and for the excerpt from *Mr. and Mrs. Pig's Evening Out*, written and illustrated by Mary Rayner, page 86; McClelland and Stewart Limited, Toronto, The Canadian Publishers for the excerpt from "Pooh and Piglet Go Hunting and Nearly Catch a Woozle" from *Winnie-the-Pooh*, by A.A. Milne, page 105; David McKay Co., Philadelphia for the excerpt from "The Death of Dermott" from *A Treasury of Irish Folklore*, published 1954, pages 44, 45; Scott Meredith Literary Agency Inc., 845 Third Avenue, New York, N.Y., 10022 for the excerpts from "Pig-Hoo-o-o-o-ey!" © 1977 the Estate of P.G. Wodehouse, reprinted by permission of the author's estate and its agents, pages 123-129; Methuen Children's Books Ltd. for the excerpt from "Pooh and Piglet Go Hunting and Nearly Catch a Woozle" from *Winnie-the-Pooh*, by A.A. Milne, page 105; William Morrow and Company for the excerpt "They pulled off their clothes and ran on faster still...." from *Pig Tale*, by Helen Oxenbury, © 1973 Helen Oxenbury, page 88; New Directions Publishing Corporation for the excerpt from *Under Milk Wood* by Dylan Thomas, © 1954 New Directions Publishing Corporation, page 119; William Nicholls for an excerpt by Brigadier General Gausson from *The Illustrated Sporting and Dramatic News*, December 22, 1933, page 35, and for "The Arrival of Black-Eyed Susan," pages 121; 122; The Osborne Collection of Early Children's Books, Toronto Public Libraries for the excerpt from *Our Dumb Neighbours or Conversations of a Father with his Children on Domestic and Other Animals*, by Thomas Jackson, M.A., published late 19th century by Patridge & Co., 39 Paternoster Row, London, page 22, *Tales of Troy and Greece*, ed. Andrew Lang, illustrated by H.J. Ford, (1907), Longman, Green & Co., 39 Paternoster Row, pages 26, 49, 50, "The Hog" from *The Juvenile Cabinet of Natural History*, London, printed and sold by R. Harrild (1820), page 78, *The Nursery Alice*, containing Twenty Coloured Enlargements from Tenniel's Illustrations to "Alice's Adventures in Wonderland" with Text Adapted to Nursery Readers by Lewis Carroll, Macmillan & Co., (1889), pages 83, 84, and "Little Dame Crump" from *Aunt Louisa's Golden Gift*, by Laura Belinda (Jewry) Valentine, Frederick Warne (1875), pages 98, 99; Oxford University Press for "The Boar's Head Carol," (arr. Martin Shaw) from *The Oxford Book of Carols*, page 47; Pan Books, London for "The Baby's Opera," pages 110, 111; Punch Publications Ltd., London for "To An Boy-Poet of the Decadence," by Owen Seaman, 1894, page 74; Random House, Inc. for the excerpts from *Cows, Pigs Wars and Witches*, by Marvin Harris, © Marvin Harris, pages 55, 56; Sterling Lord Agency, Inc. for the excerpt "Nor is there any beast..." from *The Hog Book*, by William Hedgepeth, © William Hedgepeth 1978, page 31; University of California Press for reprinting of the poem "The No-Water Boy," by Kornei Chukovsky from *A Harvest of Russian Children's Literature*, © Miriam Morton 1967, page 78; Virgil's "The Aeneid," excerpt from the translation by John Dryden in *Complete Poetical Works of Dryden*, edited by Georges R. Noyes, published by Houghton Mifflin Co., page 41; Frederick Warne & Co. Ltd. for the excerpt from *The Tale of Little Pig Robinson*, by Beatrix Potter, page 87, for the excerpt from *The Owl and the Pussy-Cat*, by Edward Lear, page 89, and for "The Story of the Three Little Pigs" from *The Golden Goose Book*, by L. Leslie Brooke, first published in 1905, pages 93, 95, 96; World's Work Ltd. for the excerpt from *Small Pig*, written and illustrated by Arnold Lobel, published in the U.K. and British Commonwealth by World's Work Ltd., page 88; Mr. Norman E. Worvill for *Dan Pig*, by Peggy Worvill, pages 90-92.

The authors wish to thank the following sources for kind permission to include illustrations in this book: Thomas Agnew & Sons Ltd. for "Circe and the Companions of Ulysses," by Briton Riviere, 1872, from *Victorian Engravings*, by Rodney K. Engen, St. Martin's Press Inc., 1975, pages 48, & 49; Atheneum Publishers for an illustration from *Mr. and Mrs. Pig's Evening Out*, written and illustrated by Mary Rayner, © 1976 Mary Rayner, page 86 (top left); and for illustrations from "As I Looked Out" from *This Little Pig-A-Wig*, chosen by Lenore Blegvad and illustrated by Erik Blegvad, © 1978 Erik Blegvad, page 89 (bottom); BBC Publications for illustrations from *Dan Pig*, by Peggy Worvill, from "Listen with Mother Stories," published by BBC Publications, pages 90, 91, 92; Ron Berg for illustrations, pages 27-29, 100-104; The Castle Howard Collection for "Girl With Pigs," by Gainsborough, page 115; Maurice Chuzeville for detail from Greek vase painting, page 42 (right); Curtis Brown Ltd. for a line illustration from E.H. Shepard, © under the Berne Convention from *Winnie-the-Pooh*, by A.A. Milne, page 105; Rodney Dennys for illustrations by A.C. Cole from *The Heraldic Imagination*, by Rodney Dennys, © 1975, The College of Arms, Queen Victoria Street, London EC4, pages 43, 44; Deborah Drew-Brook for illustrations pages 35, 64-71, 89 (top), 122; Gibson Greeting Cards Inc. for "What's the Latest Dirt?", page 78; James Gillray for "The Good Life", page 72, and "A Birch Rod or Scourge," page 73; Hamish Hamilton Limited for illustrations from "As I Looked Out" from *This Little Pig-A-Wig*, chosen by Lenore Blegvad and illustrated by Erik Blegvad, page 89 (bottom); Harper & Row Publishers Inc., for illustrations from *Small Pig*, written and illustrated by Arnold Lobel, © 1969 Arnold Lobel, page 88 (bottom); William Heinemann Ltd., Publishers for an illustration from *Pig Tale*, by Helen Oxenbury, page 88 (top); George Sherwood Hunter, 1874, for "Going to Market on a Rainy Day," page 97; Hutchinson Publishers for illustrations from *Hunting Weapons—The Arms and Armour Series*, by H.L. Blackmore, © 1971, first published by Barrie & Jenkins, page 25 from a 17th-century mᵉ. illustrated in d'Olenines, *Essai sur le Costume et les Armes des Gladiaterus* (1882), page 30 woodcut from the 1509 Venice edition of Ovid's *Metamorphoses*, page 32 (top) woodcut from *Le Livre du Roy Modus* (1486), page 34 from J. Greenwood's *Wild Sports of the World* (1862); Manya Igel Ltd., London for "Good Taste," by

William Weekes, page 74; Musée du Louvre, Paris for detail from Greek vase painting, page 42 (right); Macmillan, London and Basinstoke for illustrations from *Modern Pig-Sticking*, by A.E. Wardrop, pages 36, 38, and for an illustration from *Mr. and Mrs. Pig's Evening Out*, written and illustrated by Mary Rayner, page 86 (top left); The Maine Line Co., Rockport, Maine for "Playing Pigs," © 1980, page 86 (bottom); Peter Martin Associates, Toronto for "A Caricature History of Canadian Politics," by J.W. Bengough, page 73 (bottom right); Tony McSweeney for "Portrait of a Tattooed Pig" from *The Association of Illustrators Second Annual*, page 62; William Morrow and Company for an illustration from *Pig Tale*, by Helen Oxenbury, © 1973 Helen Oxenbury, page 88 (top); John Murray Publishers Ltd. for "Come Dancing," by Beryl Cook, page 58; Museo del Prado, Spain for "The Three Graces," by Reubens, page 75; New York State Historical Association, Cooperstown for "Ringing the Pig," by William S. Mount, page 39; Robbi Nyman for illustrations pages 56, 57, 114, 131-137; The Osborne Collection of Early Children's Books, Toronto Public Libraries for illustrations from page 7, *Stories from Old-Fashioned Children's Books*, brought together and introduced to the reader by Andrew W. Tuer, London, The Leadenhall Press Ltd., Simpkin, Marshall, Hamilton, Kent and Co. Ltd., New York, Charles Scribner's Sons (1899-1900), page 13, *Our Dumb Neighbours or Conversations of a Father with his Children on Domestic and Other Animals*, by Thomas Jackson, M.A., published late 19th century by Partridge & Co., 39 Paternoster Row, London, pages 16 (bottom left), 21, 23, 32 (bottom), 116, 120, *Stories from Old-Fashioned Children's Books*, chosen by Andrew W. Tuer, London, The Leadenhall Press Ltd., Simpkin, Marshall, Hamilton Kent and Co. Ltd., New York, Charles Scribner's Sons, page 121, *Tales of Troy and Greece*, ed. Andrew Lang, illustrated by H.J. Ford (1907), Longman, Green & Co., 39 Paternoster Row, London, page 26, "The Wild Boar" from *The Good Child's Cabinet of Natural History, "Beasts,"* Vol. 1, John Wallis, Ludgate Street, London (June 12, 1801), page 33, "Flying Pig" from *The World Turned Upside-Down*, illustrated by Wonderful Prints, W. & T. Darton, page 40, "The Hog" from The Juvenile Cabinet of Natural History, London, printed and sold by R. Harrild (1820), page 47, "Sow and Litter" from *Vere Foster's Drawing Copy Book*, No. 02 in Animal Series, illustrations by Harrison Weir, Blackie and Son, Old Bailey, (1855), page 83, *The Nursery Alice*, containing Twenty Coloured Enlargements from Tenniels Illustrations to "Alice's Adventures in Wonderland" with Text Adapted to Nursery Readers by Lewis Carroll, London, Macmillan & Co., (1889), page 84, "This Little Pig Went to Market," page 85, and *Aunt Louisa's Golden Gift*, by Laura Belinda (Jewry) Valentine, Frederick Warne (1875), pages 98-99; Derek Parks-Carter for illustrations pages 60, 61; Erik J. Peters for jacket illustration; Penguin, U.S. for "Come Dancing," by Beryl Cook, page 58; Gaston Phoebus for "Medieval Hunting Scenes," page 31; Graham Pilsworth for illustrations pages 9, 46, 76, 77, 123-129; Rijksmuseum-Stichting, Amsterdam for "Arrogance Becomes an Ass," by Dürer, from S. Brant's Narrenschiff, page 15, and for "The Hog," Ref. B. 157, by Rembrandt, page 119; Deborah Rogers Ltd., Literary Agency, London for "Come Dancing," by Beryl Cook, page 58; George Routledge & Sons, 1881, for "The Farmer's Boy," by R. Caldecott, courtesy of The Osborne Collection, Toronto, page 2; Linda Roy for illustration page 63; Saiga Publishing Company Limited for "The Old Irish Greyhound Pig," after an illustration by Richardson, © 1850 from *The Book of the Pig*, by Susan Hulme, page 20; Ronald Searle and Hope Leresche & Sayle, Literary and Dramatic Agents, London for "The Painting Pig," page 59; Ken Steacy for illustrations pages 51-54, 79, 80-81, 106-109, 138-140; Shizuye Takashima, page 142; Mrs C. R. Taylor of Santa Barbara, California for "Circe and the Pigs," by Marc Chagall, page 42; Mr Andrew Thomas of Andrew Thomas Gallery, Oxford for illustrations pages 14, 16, 17, 19, 45 and 117, for Etching by Sir Edwin Landseer, "A British Boar," (1818), page 18, for "Large Black Hog," page 19 (bottom), for Wild Boar from "The Pig," by William Youart, 1847, page 24, for "A Fat Sow," from *Marmer's Magazine*, No. 3, Vol.7 (1843), page 41, and for "Edward VII and His Pig at Smithfield," (1904), by Alex Butler, page 118; Lewis Vardey for "The Rogues Gallery," pages 10-11; Frederick Warne & Co. Ltd. for illustrations from *The Tale of Little Pig Robinson*, by Beatrix Potter, page 87, *Cecily Parsley's Nursery Rhymes*, by Beatrix Potter, page 96 (bottom), "The Little Black Pig has a Bath" from *The Art of Beatrix Potter*, page 82, and "The Story of the Three Little Pigs" from *The Golden Goose Book*, by L. Leslie Brooke, first published in 1905, pages 94, 95, 96 (top); World's Work Ltd. for illustrations from *Small Pig*, written and illustrated by Arnold Lobel, published in the U.K. and British Commonwealth by World's Work Ltd. page 88 (bottom).

Every effort has been made to ensure that permissions for all material were obtained. Those sources not formally acknowledged here will be included in all future editions of this book.